Stefan Haslinger

Cold Atoms in a Cryogenic Environment

Stefan Haslinger

Cold Atoms in a Cryogenic Environment

Technology and Engineering

Südwestdeutscher Verlag für Hochschulschriften

Impressum/Imprint (nur für Deutschland/only for Germany)
Bibliografische Information der Deutschen Nationalbibliothek: Die Deutsche Nationalbibliothek verzeichnet diese Publikation in der Deutschen Nationalbibliografie; detaillierte bibliografische Daten sind im Internet über http://dnb.d-nb.de abrufbar.
Alle in diesem Buch genannten Marken und Produktnamen unterliegen warenzeichen-, marken- oder patentrechtlichem Schutz bzw. sind Warenzeichen oder eingetragene Warenzeichen der jeweiligen Inhaber. Die Wiedergabe von Marken, Produktnamen, Gebrauchsnamen, Handelsnamen, Warenbezeichnungen u.s.w. in diesem Werk berechtigt auch ohne besondere Kennzeichnung nicht zu der Annahme, dass solche Namen im Sinne der Warenzeichen- und Markenschutzgesetzgebung als frei zu betrachten wären und daher von jedermann benutzt werden dürften.

Coverbild: www.ingimage.com

Verlag: Südwestdeutscher Verlag für Hochschulschriften GmbH & Co. KG
Heinrich-Böcking-Str. 6-8, 66121 Saarbrücken, Deutschland
Telefon +49 681 37 20 271-1, Telefax +49 681 37 20 271-0
Email: info@svh-verlag.de

Approved by: Wien, TU, Diss., 2011

Herstellung in Deutschland:
Schaltungsdienst Lange o.H.G., Berlin
Books on Demand GmbH, Norderstedt
Reha GmbH, Saarbrücken
Amazon Distribution GmbH, Leipzig
ISBN: 978-3-8381-2883-2

Imprint (only for USA, GB)
Bibliographic information published by the Deutsche Nationalbibliothek: The Deutsche Nationalbibliothek lists this publication in the Deutsche Nationalbibliografie; detailed bibliographic data are available in the Internet at http://dnb.d-nb.de.
Any brand names and product names mentioned in this book are subject to trademark, brand or patent protection and are trademarks or registered trademarks of their respective holders. The use of brand names, product names, common names, trade names, product descriptions etc. even without a particular marking in this works is in no way to be construed to mean that such names may be regarded as unrestricted in respect of trademark and brand protection legislation and could thus be used by anyone.

Cover image: www.ingimage.com

Publisher: Südwestdeutscher Verlag für Hochschulschriften GmbH & Co. KG
Heinrich-Böcking-Str. 6-8, 66121 Saarbrücken, Germany
Phone +49 681 37 20 271-1, Fax +49 681 37 20 271-0
Email: info@svh-verlag.de

Printed in the U.S.A.
Printed in the U.K. by (see last page)
ISBN: 978-3-8381-2883-2

Copyright © 2011 by the author and Südwestdeutscher Verlag für Hochschulschriften GmbH & Co. KG and licensors
All rights reserved. Saarbrücken 2011

Dissertation

Cold Atoms in a Cryogenic Environment

ausgeführt zum Zwecke der Erlangung des akademischen Grades eines
Doktors der technischen Wissenschaften unter der Leitung von

Univ. Prof. Dipl.-Ing. Dr. Hannes-Jörg Schmiedmayer
E141
Atominstitut

eingereicht an der Technischen Universität Wien
Fakultät für Physik

von

DI Stefan Haslinger
(Matrikelnummer 0025641)
Columbusgasse 12/24, 1100 Wien

Wien, am 4. April 2011 ....

Diese Dissertation haben begutachtet: ....
J. Schmiedmayer   M. Brune

- Dedicated to Elisabeth -

1 Kurzzusammenfassung

Die Idee der Quanten-Informationsverarbeitung zieht zunehmend Interesse auf sich, wobei verschiedenartige Quanten-Objekte und Quanten-bits untersucht werden, um ideale Bausteine für Quanten-Informationsverarbeitungs-Systeme zu finden. Hybride Quantensysteme sind daher als Bausteine für solche Systeme vielversprechende Objekte, da sie die spezifischen Nachteile der einzelnen Quantenobjekte ausgleichen können.

Basierend auf der Technologie supraleitender Resonatoren, sind die in den letzen Jahren intensiv untersuchten, koplanaren Mikrowellen-Leiter eine geeignete Plattform, um Photonen und Festkörper-Qubits miteinander zu verbinden. Da für solch elektrisch zugängliche, schaltkreisähnliche Kavitäts-Quanten-Geräte derzeit ein Quanten-Speicher fehlt, welcher die schnell verarbeitenden Festkörper-Quanten-Systeme mit jenen, über aussergewöhnlich lange Kohärenzzeiten verfügende, Atom-Ensembles verbindet, konzentriert sich diese Arbeit auf die technologischen Grundlagen für die Hybridisierung bisher etablierter Quantensysteme. Die Mikrowellen-Photonen welche in einem supraleitenden Hochfinesse-Mikrowellenresonator gespeichert werden, sind also eine ideale Verbindung zwischen der Atom- und der Festkörper-Quantenwelt.

Im letzten Jahrzehnt wurde die Miniaturisierung und Integration von Manipulationstechniken aus der Quantenoptik und Atomphysik, erfolgreich auf einem einzelnen Chip realisiert. Solche Atom-Chips sind zur Quantenmanipulation von ultrakalten Atomen in der Lage und bieten eine vielseitige Plattform um die Manipulationstechniken aus der Atomphysik mit Nano-Fabrikationstechniken zu verbinden.
In den letzten Jahren eröneten mehrere erfolgreich durchgeführte Experimente zur Realisierung von supraleitenden Atom-Chips in kryogenen Umgebungen, den Weg für die Integration von supraleitenden Mikrowellenresonatoren um atomare Ensembles magnetisch an Photonen in koplanaren Hochfinesse Kavitäten zu koppeln.

Diese Arbeit erörtert die Konzeption, Gestaltung und den Versuchsaufbau von zwei Ansätzen, basierend auf einer *Elektronenstrahl getriebenen Alkali-Atom-Quelle für das Laden einer Magneto-Optischen-Falle in einer kryogenen Umgebung*, und dem *magnetischen Trans-*

1. Kurzzusammenfassung

port von ultrakalten Atomen in einen Kryostat, um ein Ensemble von ^{87}Rb-Atomen innerhalb einer kryogenen Umgebung zu realisieren und in weiterer Folge an einen Mikrowellen-Resonator zu koppeln. Zuletzt werden Ergebnisse sowohl für die Elektronenstrahl getriebene Atom-Quelle, als auch für den magnetischen Transport präsentiert, bei dem letztendlich Rubidium Atome in einer supraleitenden Magnet-Falle gefangen werden.

2 Abstract

The idea of quantum information processing attracts increasingly interest, where a complex collection of quantum objects and quantum bits are employed to find the ideal building blocks for quantum information systems. Hybrid quantum systems are therefore promising objects as they countervail the particular drawbacks of single quantum objects.

Based on superconducting resonator technology, microwave coplanar waveguides provide a well suited interconnection for photons and solid-state quantum bits (qubits), extensively investigated in recent years. Since a quantum memory is presently missing in those electrical accessible circuit cavity quantum devices, connecting the fast processing in a solid sate device to the exceptional long coherence times in atomic ensembles, the presented work is focused to establish the technological foundations for the hybridization of such quantum systems. The microwave photons stored in a superconducting high finesse microwave resonator are therefore an ideal connection between the atom and the solid state quantum world.

In the last decade, the miniaturization and integration of quantum optics and atomic physics manipulation techniques on to a single chip was successfully established. Such atom chips are capable of detailed quantum manipulation of ultra-cold atoms and provide a versatile platform to combine the manipulation techniques from atomic physics with the capability of nano-fabrication.
In recent years several experiments succeeded in realization of superconducting atom chips in cryogenic environments which opens the road for integrating super-conductive microwave resonators to magnetically couple an atomic ensemble to photons stored in the coplanar high finesse cavity.

This thesis presents the concept, design and experimental setup of two approaches to establish an atomic ensemble of ^{87}Rb atoms inside a cryogenic environment, based on an *Electron beam driven alkali metal atom source for loading a magneto optical trap in a cryogenic environment*, and a *Magnetic transport of ultra-cold atoms into a cryostat*. Results are presented, both for the electron beam driven atom source, and the magnetic transport, were an ensemble of rubidium atoms is finally trapped in a superconducting magnetic trap.

3 Abreviations

general

QED	Quantum Electrodynamics
CQED	Cavity Quantum Electrodynamics
MOT	Magneto-Optical Trap
TOP	Time-Orbiting Potential
QUIC	Quadrupole-Ioe-Configuration
BEC	Bose-Einstein Condensate
NV	Nitrogen Vacancy

technical

OFHC	Oxygen-Free High-Conductivity
UHV	Ultra High Vacuum
QMA	Quadrupole Mass Analyzer
CCS	Carbon Ceramic Sensor
AMD	Alkali Metal Dispenser
TTL	Transistor-Transistor-Logic
RF	Radio Frequency
MW	Microwave Frequency
FWHM	Full Width Half Maximum
TOF	Time of Flight
RMS	Root Mean Square
EM	Electron Microscopy
TEM	Transmission Electron Microscopy

optical

AOM	Acusto-Optical-Modulator
EOM	Electro Optical Modulator
FM	Frequency Modulation
FO	Frequency Oset
WM	Wavelength Modulation
PD	Photodiode
APD	Avalanche Photodiode
VCO	Voltage Controlled Oscillator
PBS	Polarizing Beam-Splitter
NPBS	Non-Polarizing Beam-Splitter
TA	Tapered Amplifier
DFB	Distributed Feed-Back
CCD	Charge Coupled Device

methodical

WDX	Wavelength Dispersive X-ray
EDX	Energy Dispersive X-ray
AES	Auger Electron Microscopy
ESD	Electron Stimulated Desorption
PSD	Photon Stimulated Desorption
EBPVD	E-Beam Physical Vapor Deposition
LIAD	Light Induced Atomic Desorption
SEM	Secondary Electron Microscopy
STEM	Scanning Tunneling-EM

Contents

1. **Kurzzusammenfassung**   iii

2. **Abstract**   v

3. **Abreviations**   vii

Contents   1

4. **Preface**   7
   - 4.1. Terminology 7
   - 4.2. Foundations 8
     - 4.2.1. Atom chips as versatile platform for cold atom experiments 8
     - 4.2.2. Superconducting Circuit-CQED Devices 9
     - 4.2.3. Superconducting Atom Chips 9
     - 4.2.4. Hybrid Quantum Systems 9
     - 4.2.5. The Quantum-Interconnect Project: Motivation for this thesis 9
   - 4.3. This thesis: Two concepts for cold ^{87}Rb atoms in a cryogenic environment . . 10

5. **Theory**   13
   - 5.1. Electron-bulk interaction with solids 13
     - 5.1.1. Energy loss of electrons 14
     - 5.1.2. Interaction volume 15
     - 5.1.3. Absorption coeficient of electrons 16
     - 5.1.4. Mean free path in solids 17
   - 5.2. Electrons in electric fields 17
     - 5.2.1. Field emission in high electric fields 18
     - 5.2.2. Electron optics 18
   - 5.3. Laser cooling and magnetic trapping of atoms 20
     - 5.3.1. The linear Zeeman-shift 21
     - 5.3.2. Magneto-optical trap and optical molasses 21

Contents

       5.3.3. Magnetic traps for neutral atoms 23
       5.3.4. Superconducting micro-traps 25
  5.4. Interactions, collisions and loss-rates in magnetic traps for cold atoms 26

I. Electron-beam driven atom source for cryogenic environments   33

6. Concept of an electron-beam driven atom source   35
  6.1. Motivation for modeling the desorption of atoms based on EBPVD 36
  6.2. Electron beam impinging a surface 36
       6.2.1. Penetration depth of electrons 37
       6.2.2. A 2D surface model for heating a thin film 39
  6.3. Evaporation model and crucial parameter 42
       6.3.1. Simulation of the eective spot size 45
       6.3.2. Temperature distribution of atoms 46
       6.3.3. Evaporation of particles - Mass flow calculation 48

7. Design considerations for an electron beam loaded MOT   51
  7.1. General experimental considerations 51
  7.2. Electron-beam preparation 52
       7.2.1. Electron source 52
       7.2.2. Focusing lens system 54
       7.2.3. Deflection of the e-beam 55
       7.2.4. Simulation of the electron source/lens-system 55
  7.3. Electrons vs. a magneto-optic trap 56
       7.3.1. Beam perturbation 56
       7.3.2. Electron impact ionization 58
  7.4. Rubidium Target 60
       7.4.1. Choice of the target 60
       7.4.2. Target consistency 61

8. Setup of an electron-beam loaded MOT   63
  8.1. Laser system and MOT optics 63
  8.2. Electron-gun 64
       8.2.1. Preparation of the field emission tips 64
       8.2.2. Emission properties 64
       8.2.3. Focusing electrodes 65
       8.2.4. Deflection plates 65
       8.2.5. Mounted E-gun 66
  8.3. E-gun targets 68
       8.3.1. Nitrogen cooled targets 68
       8.3.2. Standard targets 68
       8.3.3. Phosphor-screen 69
  8.4. Vacuum-system for the e-beam MOT 69
  8.5. Experimental setup 69

9. Results: The electron-beam-MOT — 73
9.1. High efficient electron-gun 73
9.2. Cold atom source 74
9.3. Model for Electron Stimulated Desorption 78

II. Magnetic transport of cold atoms into a cryogenic environment — 83

10. Concept of a magnetic transport for cold atoms — 85
10.1. Demands for a magnetic transport line 87
10.2. Basic design considerations 88
10.3. Simulation of the magnetic transport 89
    10.3.1. Coil configuration 91
    10.3.2. Calculations of the current 91

11. Setup of a magnetic transport line at room temperature — 99
11.1. Laser system 99
    11.1.1. Locking Techniques 100
    11.1.2. Cooler-Laser and amplifier 102
    11.1.3. Repumper-Laser 104
    11.1.4. Light conditioning 104
11.2. Setup of the lower transport-chamber 106
    11.2.1. The vacuum-chamber 106
    11.2.2. The transport line 107
    11.2.3. MOT-optics 107
11.3. Coil temperature control system 108
11.4. Lower Imaging system 110
11.5. Experimental setup 110
11.6. Experimental cycle and transport schemes 112

12. Design of a 4K cryo-system — 113
12.1. Cryogenic cooling systems 113
    12.1.1. Bath cryostats 114
    12.1.2. Closed Cycle Refridgerants 114
    12.1.3. ARS closed cycle cryo-head 115
    12.1.4. Thermal heat budget 115
12.2. Design goals for a cryogenic cold atom experiment 116
    12.2.1. Superconducting vertical magnetic transport line at 4K 117
    12.2.2. Design of a superconducting quadrupole-Ioe trap 117
    12.2.3. Superconducting micro-trap 119
    12.2.4. Atom chip mounting 119
    12.2.5. Current wires down to 4K 120
    12.2.6. Good optical access 121
    12.2.7. Fast switching of SC coils 122
12.3. 4K experimental stage 123
    12.3.1. Design considerations for the 4K stage 123
    12.3.2. Connectors 124

12.3.3. Coil-mountings	124
12.4. Thermal shielding	124
12.4.1. Anchoring of windows at low temperatures	124
12.4.2. Design considerations for a cryo shield	124
12.4.3. Material and heat conduction	125
12.5. Cryostat vacuum-chamber	125

13. Setup of a 4K cryo-system   127
- 13.1. The Cold finger and the Vacuum system 127
- 13.2. The radiation shield 127
- 13.3. Inner life of the cryostat 129
- 13.4. The superconducting coils 134
  - 13.4.1. The vertical transport coils 134
  - 13.4.2. The QUIC- and super-Ioe-coils 135
  - 13.4.3. Solder contacts for the SC coils 136
- 13.5. 4K system and the transport section 136
  - 13.5.1. Upper imaging system 138

14. Results: Cold atoms inside a 4K-cryogenic machine   139
- 14.1. Transport of atoms into the cryostat 139
  - 14.1.1. SC vertical transport scheme 139
  - 14.1.2. Imaging the atoms in the cryostat 140
  - 14.1.3. Spatial oscillations during free fall 141
- 14.2. Optimization of the transport 143
  - 14.2.1. Lower magnetic trap 143
  - 14.2.2. Horizontal transport 144
  - 14.2.3. Vertical transport 145
- 14.3. SC quadrupole trap 147
  - 14.3.1. A large superconducting quadrupole trap 147
  - 14.3.2. Lifetime in the cryogenic environment 148
  - 14.3.3. In-situ tomography of the sc-quadrupole trap 149
- 14.4. A superconducting Ioe-Pritchard-like trap 150

15. Towards: Cold atoms in a superconducting micro-trap   153
- 15.1. Re-loading atoms in a superconducting micro-trap 153
  - 15.1.1. The initial QUIC-trap 154
  - 15.1.2. The final z-trap 154
  - 15.1.3. Reloading to a micro-trap 155
- 15.2. Trapping atoms in a superconducting chip-based micro-trap 158

III. Conclusion and Outlook   161

16. Conclusion   163

17. Outlook   165

IV. Addenda — 171

18. Sincerce thanks are given to... — 173

List of Figures — 175

List of Tables — 179

Bibliography — 181

V. Appendix — 197

A. Experimental — 199
    A.1. Performance and setup of the ARS closed cycle cryo-head 199
    A.2. Windows at 50K 201
    A.3. The challenge of SC coils 203
        A.3.1. Winding the coil 203
        A.3.2. Dissipation at the solder contact 204
        A.3.3. Radiation input from the windows 206
    A.4. Transport stability 208
    A.5. Lifetime issues in the cryostat 210
        A.5.1. Improved dierential pumping stage 211
        A.5.2. 4K cryogenic pump using charcoal 211
        A.5.3. Pulsed cryostat 211
    A.6. Eddy current issues and oscillating magnetic fields 212
    A.7. Ingredients to build up the 4K environment 214
    A.8. Method of detection: Absorption Imaging of cold atoms 215

B. Electronics — 219

C. Experimental control — 225

D. Material Properties — 229

E. Rubidium data — 233

F. Physical constants — 235

G. Construcion Drawings — 237

4 Preface

Since the postulation of quantization of light by Max Planck in 1899, and Einstein's explanation of the photoelectric eect 1905, the general validity of the idea of quantization finally led to the realization of the laser. In the 1960s additionally a lot of work was done on quantum electrodynamics (QED), the relativistic quantum field theory of electrodynamics. QED celebrated its success as it is the first theory achieving full agreement between quantum mechanics and special relativity, and provided complete account for light and matter interaction.

As atoms are considered as quantum mechanical oscillators with discrete energy levels, with transitions driven by absorption and emission of photons, the laser opened the road for extensive investigations of the atomic energy spectrum. Hence, the interaction of light and matter was studied, with strong focus on the mechanical forces appearing between photons and atoms. With trapping of atoms, either by magnetic forces, or radiation pressure and the Doppler cooling of those samples, the road opened towards the cooling of atomic ensembles and finally achieving Bose-Einstein condensation in the mid 1990s [1, 2] which was theoretically proposed already in the mid 1920s [3, 4].

Expanded to the interaction of matter with light confined in a cavity, CQED experienced a further extension based on micro-fabrication. Hence Circuit-CQED, provides three great experimental advantages: a) it provides extremely high field-strength densities as the transversal cavity confinement is extremely high, b) it allows for electrical manipulation as the cavity operates in the GHz-frequency-regime, and c) the cavity experimentally allows for a huge variety of quantum objects to be implemented. This covers atoms, molecules and ions, arbitrary spin ensembles, NV-color centers, and artificial *solid-state atoms*, to study their interaction and searching for realization of quantum information processing.

4.1. Terminology

Laser Cooling of atoms was one of the greatest achievements in experimental atomic- and quantum-physics in the past two decades and has become a standard technique preparing ensembles of ultra-cold atoms. Slightly counterintuitive, laser light is used to decrease the mean velocity of an atomic ensemble, and hence slowing them down. It relies on directed

4. Preface

photon absorption while emission occurs isotropic.

Magnetic Trapping of atoms can be realized in magnetic gradient fields, trapping neutral particles with a magnetic moment. In combination with the development of laser cooling in the early 90ties, cold trapped atoms allowed for novel experiments studying the interaction of light and cold quantum gases opening the new field of atom optics.

Atom Chips provide a versatile platform for manipulating ensembles of cold, trapped atoms. Chip-like substrates with micro- and nano-fabricated current carrying wires allow for strong magnetic micro-traps, multilayer designs and additional allow optical fibers, permanent magnets and prisms to be embedded, realizing a quantum lab on the chip.

Cryogenic Environments are experimentally realized since liquid helium and nitrogen are known for more than a century. Modern cryogenic systems have became dry- and closed-cycle-systems, capable of challenging experiments far beyond 4K. Since atom chips have gone superconducting in recent years, cold atom experiments realizing chip-traps in cryogenic environments have been upset.

Circuit-CQED is known as a quasi-2D diminution of the classical 3D-geometric cavity, archetype for CQED systems. It transcribes an amazing experimental extension for CQED towards electrical circuits. These coplanar transmission line resonators situated in the microwave-frequency regime oer a platform to integrated solid-state quantum devices for quantum information processing.

Hybrid Quantum Systems can be realized by coupling two dierent quantized quasi-two-level systems such as solid-state quantum bits (qubits) and cold atomic ensembles. Based on a Circuit-CQED scheme, a superconducting atom chip can be an ideal platform, interconnecting those dierent quantum systems.

4.2. Foundations

This introduction briefly describes tools, devices and systems defining the framework in which my work is embedded. Covering the two worlds of atom chips[1] and superconducting quantum devices, the last paragraph sketches the *quantum interconnection project*.

4.2.1. Atom chips as versatile platform for cold atom experiments

Since atom cooling and trapping techniques developed very fast in the late 90ties, interactions of cold atoms and charged wires were studied [5], which consequential lead to the implementation of micro-fabricated wire traps onto substrates [6, 7, 8, 9], so called atom chips. This miniaturization opened the road for atom chips to establish as a versatile platform for the manipulation of trapped and cold ensembles of atoms. Capable of double-layer structures and wires down to the sub μm-regime, wires can withstand current densities up to $10^{12} A/m^2$ exceeding comparable current densities in free-standing wires by orders of magnitude [10]. The

---
[1] Atom chip: A micro-fabricated, planar device which allows to magnetically trap and cool cold atoms.

resulting strong magnetic micro-traps therefore allows to achieve Bose-Einstein condensation (BEC) in trapping geometries much smaller than known from the first experiments [1, 2]. Dierent experiments exhibited atom chips as a ideal tool sensing electric and magnetic fields with a BEC [11] or detecting single atoms with integrated photon detectors [12, 13].

4.2.2. Superconducting Circuit-CQED Devices

As demonstrated in recent years, the fabrication of superconducting devices for Circuit-CQED [14], capable of coupling a superconducting quantum two-level system to a single microwave photon [15, 16], both relies and demands micro-fabricated coplanar waveguide structures [17]. Since experiments trapping atomic ensembles on superconducting atom chips[18, 19] were performed, superconducting Circuit-CQED devices and atom chips can be thought to combine atomic- and solid-state physics.

4.2.3. Superconducting Atom Chips

Since the first proposal for a superconducting atom chip [20], recent experiments with atomic ensembles in superconducting micro-traps [21] and atom chip based superconducting micro-traps [18, 19] investigated the influence of the *Meissner-eect* [21, 22] and lifetimes close to superconducting Nb and MgB_2 surfaces [23, 24, 25] at $T << 77K$, finally achieving Bose-Einstein condensation on a superconducting atom chip [26].
In addition experiments using superconducting atom chips even at higher temperatures ($T < 87K$) were setup [27, 28, 29] mainly based on the high-T_C superconductor $YBa_2Cu_3O_{7-x}$.

4.2.4. Hybrid Quantum Systems

The interconnection between atom-like quantum systems and solid-state qubits [30], especially cold quantum gases and superconducting resonator based solid-state qubits [31, 32, 33, 34], often with countervailing advantages, at least aims to outrange the disadvantages, to combine the best of two worlds. Solid-state quantum systems therefore provide their properties to be controlled in a wide range by micro-fabrication techniques. On the other hand, with the realization, that a single excitation in a collective state in a BEC or an ultra-cold atomic ensembles is a very robust quantum bit [35], atomic ensembles attracted interest as well.
Therefore the science of superconducting quantum devices is proposed to play a major role in the architecture of quantum information processing and quantum computing with atomic ensembles [34].

4.2.5. The Quantum-Interconnect Project: Motivation for this thesis

Successful coupling of quantum systems to superconducting resonators in the microwave frequency regime were recently demonstrated using rare earth ion-doped crystals [36], electron spin-ensembles [37], and nitrogen-vacancy centers in diamonds [38]. A superconducting atom chip would therefore provide a flexible platform to integrate a planar microwave-resonator on the one hand, and allows for manipulation and trapping of ultra-cold atomic ensemble on the other hand. To open the scientific road towards such hybrid quantum systems with an atomic ensemble coupled to a solid-state device, cold ^{87}Rb atoms have to be established inside a cryogenic environment suitable for the implementation of a microwave cavity and superconducting solid-state qubits.

4. Preface

The microwave photons, a standing electromagnetic wave in the superconducting resonator would couple magnetically via a hyperfine transition to the atomic ensemble demanding the rubidium atoms to be very close to the surface [31]. The best choice for the coupling qubit states in such a hybrid quantum system, would be the trappable clock states $|F=1, m_F=-1\rangle$ and $|F=2, m_F=1\rangle$ where very long coherence times $> 1s$ were demonstrated for hyperfine excitations on atom chips even at close proximity to the surface [39]. Nevertheless this demands a two-photon transition to satisfy the selection-rules, introducing a virtual level and driving the transition $|F=1, m_F=-1\rangle \rightarrow |F=2, virtual\rangle \rightarrow |F=2, m_F=1\rangle$ with a cavity MW-photon and a RF-photon.

According to recent calculations [31] using a coplanar waveguide geometry which is optimized for maximal magnetic field and holding atoms several few μm from the surface, the coupling to a single atom will reach $\approx 40Hz$. An ensemble of 10^6 atoms would therefore translate this coupling to a collective Rabi-frequency of up to 40 kHz for creating a collective spin wave excitation in the ensemble, presenting a robust Dicke-state for a single excitation written into the cold cloud of atoms.

4.3. This thesis: Two concepts for cold ^{87}Rb atoms in a cryogenic environment

The aim of realizing a robust quantum memory for quantum information processing devices as previously mentioned, is therefore the framework in which my thesis is embedded. To establish the experimental foundations for this challenging goal, I describe in my thesis two novel methods to achieve cold atoms in a cryogenic environment[2].

While the letter restricts realizations to be compatible with the limited cooling power in cryogenic systems, the design goals for the experiment cover more demands: a) A high atom number in the cryogenic environment[3], b) space for the implementation of a superconducting atom chip, which allows manipulation of the atoms in a superconducting micro-trap, and c) which allows for the implementation of a superconducting resonator, with d) almost no light facing the latter[4], as a resonator-shift would influence the two-level transition crucially.

This finally leads to two possible concepts. Either a *source for cold atoms inside a cryogenic environment*, including a magneto-optical Trap (MOT) to prepare an ensemble of cold atoms, or a *transport scheme of atoms inside a cryogenic environment*.

Typical sources for alkali metal atoms as they are used in BEC-experiments[5], are among others, ovens [46] releasing alkali metal atoms feeding Zeeman-slowers [47] or alkali metal dis-

---

[2] I already mentioned superconducting micro-traps for trapping ensemble of atoms, most of them based on atom chips and realized in recent years. Nevertheless those traps did not for the first time realize a superconducting trap for cold atoms as this was already done in 1995 for the first time [40]

[3] To increase the single atom-photon coupling at given parameter this is just possible by increasing the number of atoms N.

[4] According to the intrinsic properties of superconducting microwave resonators [17], recent investigations in our lab indeed have shown ineligible frequency shifts from the resonance frequency $\omega_0 = 6.83GHz$ if the resonator faces even several nW of light power at certain frequencies

[5] For detailed references the reader is guided to [41, 42, 43, 44, 45]

4.3. This thesis: Two concepts for cold ^{87}Rb atoms in a cryogenic environment

penser (AMD)[6] generating a vapor using resistive heating[7] where the MOT is loaded from. As these sources does not comply with the demands in a cryogenic system, a novel low power atom source is needed, even if it is not clear if it can provide a sufficient amount of trappable ^{87}Rb atoms.

State of the art experiments with ultra-cold atoms on superconducting atom chips, therefore use rather sophisticated transport schemes [18, 19, 25, 23, 24, 26]. The different implementations are realized by a movable magnetic trap [19] transferring the atoms into the cryogenic environment, optical tweezers [21], or pushing a beam of atoms into the cryostat, catching the atoms in an implemented MOT [26]. Above mentioned transport schemes do not satisfy all demands realizing a prospective implementation of super-conductive solid-state quantum devices. Therefore a transport scheme without the use of light, which relies on an all magnetic transfer with static coils is realized.

The two concepts to establish cold atoms in a cryogenic environment covers the following topics:

Field emission electron-beam source which is capable of high beam currents at moderate emission and acceleration voltages. It allows the electron beam properties to be tuned over a wide range, featuring a tunable beam power, and hence providing an extraordinary flexible heat source even compatible with the demands for cryogenic systems.

Electron beam driven atom source for loading a MOT, using a flexible, fully tunable electron source which can desorb alkali metals from a cold surface to subsequently trap them in a MOT. As the electron energy is almost instantaneous deposited at the target, this atom source could allow for huge loading rates trapping the atoms in a MOT.

Magnetic transport into a cryogenic system allows to transfer a cold, trapped ensemble of ^{87}Rb atoms over a large distance. It relies on a well established technique which is extended by a vertical transport with partly normal conducting and superconducting coils, suitable to reload atoms adiabatically into chip-based micro traps.

4K cryogenic system capable for a superconducting atom chip-experiment is setup from the scratch and provides the foundation for the implementation of a superconducting resonator coupled to cold ^{87}Rb atoms. It offers experimental flexibility and combines the world of cryogenics with ultra-cold atom physics. This experimental setup would thereby be the test bench for the magnetic transport system and the implementations in the cryogenic environment to adapt the experiment even to temperatures much below 4K.

The thesis consists of two parts and is organized as follows:

- Chapter 4 is about the theory behind electron beams and their interaction with bulk materials, as well as the standard laser-cooling and trapping techniques established for cold atom physics covering magnetic-traps and magnetic micro-traps. Additionally it describes the basics on collisions and loss-rates in cold quantum gases.

---
[6]SAES Getters S.p.A, 20151 Milano, Italy
[7]alkali metal atoms are released via the compound is chemically reduced by resistive heating of the crucible

4. Preface

- **Part I** describes the electron-beam driven atom source for cryogenic environments.

- Chapter 5 discusses the concept of an electron-beam driven atom source and addresses the question how this novel atom source can be realized.

- Chapter 6 considers the design demands for a setup where an electron beam is used to load a MOT.

- Chapter 7 gives the setup of such a device including the development of a high current field emission electron gun.

- Chapter 8 shows the results achieved with this electron-beam MOT and comes up with a simple model to describe the relying eect of desorbing rubidium atoms.

- **Part II** describes a magnetic transport of cold atoms into a cryogenic environment using a setup of non-moving magnetic coils only, with the atoms finally trapped in a superconducting, macroscopic quadrupole-Ioe trap.

- Chapter 9 discusses the concept of magnetic transport schemes and the resulting demands to be experimentally realized.

- Chapter 10 is about the setup of a horizontal room-temperature transport scheme extended by the first realization of a vertical magnetic transport with static coils.

- Chapter 11 describes the design of the 4K cryogenic system necessary to setup the desired experiment

- Chapter 12 shows the setup of the 4K environment with the superconducting vertical transfer coils.

- Chapter 13 presents the results of the achieved transport setup and detailed characterization and additionally presents cold atoms trapped in the superconducting traps.

- Chapter 14 continues with calculations of a reloading scheme to bring atoms close to the surface of a superconducting atom chip, and to trap them in a chip-based micro-trap.

- Chapter 15 summarizes and gives an outlook of future projects at this experiment.

- The Appendix gives detailed information about some experimental issues and questions which are worth to be addressed.

5 Theory

Desorption of rubidium atoms from a surface with subsequent trapping and cooling of the neutral atoms, is based on an electron beam impinging a proper target. Hence the development of an electron gun for preparing a beam, capable of a certain power density, leads to a combination of methods known from applications in physical analytics and electron beam induced coating techniques. This ranges from the choice of the proper electron lens system for focusing, to the choice for the right electron emitter, combining technological ingredients which are in this certain constellation rather uncommon [1]. Furthermore electron-bulk interactions are usually known from analytical techniques for element analysis or microscopy, but were also under investigation for electron beam induced deposition [48].

This section therefore focus especially on electron beam - bulk interactions, as a basis for the following design and conceptual considerations, and to clarify in section (9) on which eects the observed results rely.

Furthermore the chapter will briefly introduce the background knowledge relevant for laser-cooling and magnetic trapping, which has become a standard technique in the past two decades.

5.1. Electron-bulk interaction with solids

Interactions of electrons with materials underlay quite well understood models since these interactions are crucial for a whole zoo of analytical techniques. Fig.(5.1) therefore gives an overview which conversion products occur when an electron beam hits a surface, and exhibits the analytical methods, exploiting the underlying process.

Photons in the x-ray energy range are to be analyzed in wavelength dispersive x-ray spectroscopy (WDX) and energy dispersive x-ray spectroscopy (EDX), both used within secondary electron microscopy (SEM) and scanning tunneling electron microscopy (STEM). Further more secondary electrons in addition with backscattered electrons are used in SEM for image generation, whereas also Auger electron spectroscopy (AES) uses electrons to generate pictures. Cathode-luminescence uses emitted light in the visible range, and specimen current

---
[1] As electrostatic lens-systems are usually used in electron microscopy, huge industrial electron beam evaporation machines use huge currents and magnetic focusing systems

5. Theory

imaging relies on the use of absorbed electrons, nothing else than measuring the target current.
All those interactions of electrons with bulk material intensively known from analytics, as depicted in Fig.(5.1), are either elastic or inelastic scattering events where both energy and momentum is transferred in dierent ways to bulk-atoms and therefore into the material.
The illustrated interaction-ranges refers to a Cu-bulk and are approximate values common in a microscopy setup with 20keV electrons [49]. The related analytical methods are written in black above the corresponding particles, either electrons or photons, detected from the surface.

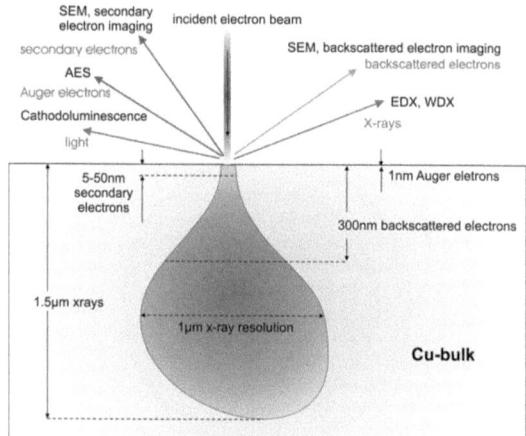

Figure 5.1.: This figure shows analytical techniques for probing a surface with an electron beam, and analyzing emitted/scattered electrons/photons from the specimen to learn about electron-bulk interaction and the specimen. It gives an overview over dierent processes and in which spatial region they occur for an illustrated example of a Cu-bulk with an incident electron beam with 20keV, regarding to [49].

5.1.1. Energy loss of electrons

Using an electron beam with energies from 1-6keV, energy conversion into photon radiation in the x-ray band is neglected [2], as well as the strong dependence on the band structure of electron-electron scattering in free-electron like metals such as alkali metals [50].
Inelastic collisions commonly occur as the so called *electronic stopping* of electrons which are due to the collisions with atomic electrons, where electrons are excited or ejected ending with a small momentum transfer but a large energy loss of the incident electron [51]. With higher electron energy this process is a single atom-electron interaction as the *de Broglie-wavelength*

---
[2] As even for continuous x-ray radiation (Bremsstrahlung) *Kramers law* gives a hyperbolic intensity decay with energy, the energy loss of electrons in ^{87}Rb due to x-ray radiation in the used energy region is negligible

gets smaller with increasing energy. Another inelastic process namely plasmonic excitations can be neglected, because of the small energy loss compared to that in atomic collisions, as well as phonon excitations which are in the order of $k_B T \approx 25 meV$.

On the other hand, transfer of energy and momentum arise from nearly elastic collisions of electrons with nuclei. This nuclear scattering therefore is the reason for the backscattered electrons [51]. If scattering of electrons by a pure Coulomb-potential occurs, also photons are emitted [50], where most of them are in the infrared region.
To complete the picture of processes, the electrons are supposed to travel straight into the target, loosing energy due to inelastic, electronic collisions, producing auger- and/or secondary electrons and are deflected or even backscattered by nuclear collisions.

5.1.2. Interaction volume

For an accurate calculation of electron beam evaporation as a possible concept underlying an electron beam driven atom source, it is crucial not only to know about the electron-bulk interactions in general, but also to be able to estimate the interaction volume for the conversion of electron energy into heat. Since it is known from analytics, that certain processes take place in different bulk depths, as depicted in Fig.(5.1), where Auger electrons and secondary electrons are spatially related to the surface region, and backscattered electrons origin more from the bulk, an accurate model has to consider more than one interaction-process.
Critical parameters defining the interaction volume are therefore the beam energy, the atomic number Z of the bulk material, and the incident angle of the electron beam . As the electron range can approximately be corrected by $R(\ ) = R_0\ cos(\ )$ [49], the electron beam energy defines the interaction intensity via the correlated cross-sections.

In principal, there are two ways to measure the electron penetration range. For electron energies below 100eV, the usual *Bethe-Bloch* formalism is inadequate for calculating the electron energy loss in a solid, and an approach using the dielectric response of the material can be used [52].
A more general approach delivers the *Kanaya-Okayama* range [51], which comes from theory explaining phenomena connected with electron penetration. The interaction volume is estimated through a sphere which gives the maximum range for penetration and energy dissipation shown in Fig.(5.2), as the electron range in solids is a measure of the straight-line penetration distance [53].
In the following the penetration depth is estimated using the empirical *Kanaya-Okayama* range, comparing it with a numerical method.

$$D_{KO} = 0.0276\ M \frac{U^{1.67}}{Z^{0.889}} \qquad (5.1)$$

The penetration depth $[D_{KO}] = 1\mu m$ or *Kanaya-Okayama* range is given in Eq.(5.1) where M is the atomic weight, U the kinetic potential of the electrons in [kV], Z the atomic number and the density of the target material in $[g]/[cm^3]$.

5. Theory

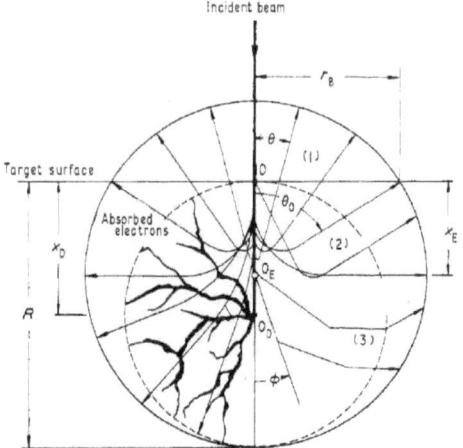

Figure 5.2.: This figure, taken from [51], illustrates how the KO-range is defined by the maximum penetration depth. It shows a modified diffusion model of electron beam penetration in a target. R is the maximum range, x_D the diffusion depth, x_E the maximum energy dissipation depth, and r_B the backscattering range with $\tan(\theta_0) = r_B/x_E$; (1), (2) and (3) refers to the number of times the electrons are deflected.

5.1.3. Absorption coefficient of electrons

Another parameter that is provided by the *Kanaya-Okayama* model is the absorption coefficient for electrons hitting a target. The coefficient can be written as

$$\eta_A = 1 - (\eta_T + \eta_B) \qquad (5.2)$$

whereas η_T and η_B are the transmission and the backscattering coefficient. Assuming the bulk to be thick enough, hence $\eta_T = 0$ and the absorption coefficient is directly related to backscattering where η_B looks like

$$\eta_B = \frac{6}{5}\int_0^y \frac{B}{1-y^{7/6}} exp^{-\frac{\gamma_B y}{1-y}} dy - \frac{6}{5}\frac{1}{2^{5/6}}\left(1 - e^{-\frac{\gamma_B y}{1-y}}\right) \qquad (5.3)$$

Here the reduced depth y enters as $y = x/R$ with the maximum range R, and x as the depth variable. The parameter η_B is related to the atomic number Z only and accounts for diffusion of returning electrons and for multiple electronic collisions, and follows $\eta_B \propto Z^{2/3}$. Therefore Fig.(5.3) shows the fractional backscattering, regarding to Eq.(5.3) for various atomic numbers Z. Assuming that even for alkali metals the backscattering cross section is not completely different, η_B is assumed to be not much bigger than 0.4 for the most disadvantageous cases

[54]. As the main contribution for electron energy dissipation in the bulk will be due to inelastic electronic stopping, producing auger- and/or secondary-electrons, the absorption coe cient $_A$ will be a measure of how much initial energy of the electron beam has been converted into heat.

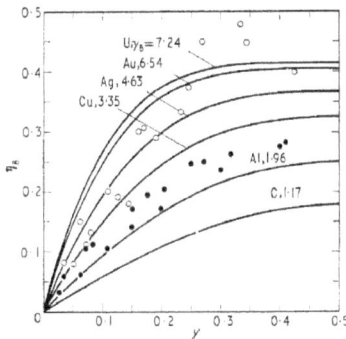

Figure 5.3.: This figure was taken from [51] to illustrate fractional backscattering. It shows e ciencies $_B$ from several targets as a function of reduced depth from the *Kanaya-Okayama* model, compared with experimental results from *Cosslett and Thomas* (1965) for Cu(•) and Au(○) at 5-20kV

5.1.4. Mean free path in solids

As the mean free path of electrons is the driving parameter entering the interaction cross-sections, this length was intensively investigated for di erent energy regions [50], respectively for alkali metals [55]. In fact, for high energies (100eV and higher) the electron solid interaction can accurately be described using atomic models [56, 57] where the mean free path is than given due to the *Bethe* formula.

Even as the Bethe formula, or the *Kanaya-Okayama* range are reliable methods to determine the penetration depth in the desired energy region of $1 - 6 keV$, the investigations of the mean free path lead to a more practical tool, namely using a Monte-Carlo Simulation to stochastically calculate single-electron trajectories in bulk materials as shown in section (6.2.1).

5.2. Electrons in electric fields

The following section concerns both, the extraction of electrons from a solid surface in high electric fields, and the basics of electron optics describing the motion of accelerated electrons in electric potentials.

5. Theory

5.2.1. Field emission in high electric fields

Electron field emission, also known as cold field emission, occurs due to the presence of high electric fields and describes the emission of electrons from solids, mostly metals, even nowadays carbon nano-tube field emitters became popular [58, 59]. In contrast, thermionic field emission, even occurs at smaller or even without electric fields if the electrons carry enough energy to overcome the potential barrier [60], added by thermal heating of the emitter. The effect of field emission, basically relies on the tunneling of electrons and was discovered experimentally more than 80 years ago, theoretically explained by the *Fowler-Nordheim* theory [61]. The field emission current density for a metal surface with the applied electric field E is given by

$$j(E) \propto E^2 \times e^{-\frac{c}{E}} \qquad (5.4)$$

and strongly depends on the field strength which enables the tunneling process. Additionally, this process relies on the work function ϕ representing the energy, which must be overcome by the electron to leave the solid. Hence the emitted current density, reads in more detail as

$$j(E) = \frac{A \times E^2}{\phi} \times e^{-\frac{B \times \phi^{3/2}}{E}} \qquad (5.5)$$

usually with A and B directly originating from the derivation of an electron in a potential with a tunneling barrier that drops linear with distance.

Depending on the tip geometry and the material properties, coefficients A and B can be experimentally derived, fitting the measured current and voltage, if the tip is assumed to be point-like [62].

According to field emission studies [63], Eq.(5.5) can be written more convenient for $[j] = A/cm^2$, $[E] = V/cm$ and the work function $[\phi] = eV$, using $A = 1.54 \times 10^{-6}$ and $B = 6.83 \times 10^{+7}$. As mentioned above, the tip geometry is crucial as the exponential behavior leads to emission in a region of high electric fields, either maintained by high applied voltages, or a small emission tip. Therefore the constants A and B should be derived by fitting measurements for an accurate consideration.

Several studies [60] of cold field emission into vacuum address questions concerning the preparation of tips and the achievable geometries [64], as well as their lifetime-cycle [65] over many experimental runs or a certain time.

5.2.2. Electron optics

The dynamics of a beam of charged particles is given by the fundamental *Maxwell's* equations for vacuum. Considering static fields, the dynamics of the systems evolve into linear-time dependent equations of motion for the particle velocities in homogeneous electric fields and relies on the Lorentz-force for electrons in magnetic fields, as found in many textbooks like [66].

In the following, I will consider three different types of interactions for electrons with electric and magnetic fields, two of them shown in Fig.(5.4).

5.2. Electrons in electric fields

a) linear electric gradient field

b) quadrupole magnetic field lines

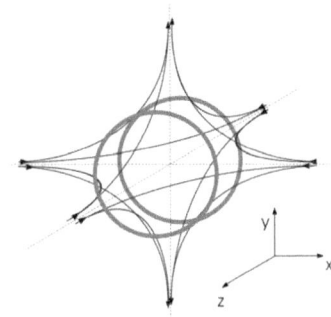

Figure 5.4.: Dierent field configurations for the movement of electrons in a) linear electrostatic gradient fields, and b) magnetic quadrupole fields generated by to anti-Helmholtz configured coils.

Electrons in a linear vector field

The non-relativistic velocity of electrons which have fallen through a electrostatic potential reads as $v_0 = \sqrt{2U\ e/m_e}$. Entering a region with length l_1 providing a linear electric field gradient as shown in Fig.(5.4) inset a), with no initial transversal velocity, the field accelerates the particle building up a velocity v_y parallel to the field lines regarding to the equation of motion.

$$F_y = m_e \frac{\partial v_y}{\partial t} = e\ E(x) \tag{5.6}$$

With a constant potential U_D and certain geometric boundary conditions the electric field $E(x) \rightarrow E(t)$ can be expressed in the inertial frame of the particle moving at constant velocity in x-direction

$$E(t) = U_D f(t) \tag{5.7}$$

where $f(x)$ and $f(t)$ depends on the electrode geometry, and $x \rightarrow t = l_1/v_0$. Applying $E(x)$ to Eq.(5.6), the electron-position transversal to the initial velocity can be calculated solving

$$y_1(t) = \int_0^t \frac{e}{m_e}\ U f(t) dt \tag{5.8}$$

Leaving the field region l_1 after t_1, the electron is preserving the transversal velocity $v_y(t_1)$, and accumulates an additional position y_2 after t_2 in a field free region. Therefore the total transversal displacement from the axis, passing a region l_1 with a transversal field changing with space, and a field free region of length l_2 cumulates as

5. Theory

$$y_{(t_1+t_2)} = v_y(t_1)t_2 + \int_0^{t_1} \frac{e}{m_e} Uf(t)dt \qquad (5.9)$$

Electrons in an arbitrary vector field

Considering an arbitrary vector field due to a non-trivial arrangement of dierent potentials in space the electron trajectories are much more di cult to calculate.
Nevertheless this can numerically be solved using dierent common software packages which calculate electron trajectories in arbitrary electric potentials such as SIMION and CPO[3], or COMSOL[4].

Solving $\Delta = 0$ in vacuum, the electric potential can be calculated in the complete space just defined by the initial boundary conditions. Neglecting the space charge of multiple particle beams and interactions in low intensity beams, the trajectories and the beam physics of several electrons can be simulated even in more complicated setups.

Electrons in magneto-static fields

Based on the Lorentz force, the trajectory of an electron in a static magnetic field can be calculated following

$$F = q \ (v \otimes B) \qquad (5.10)$$

and noticing that the particle motion however evolves into a 3D trajectory even if the initial velocity is restricted to one dimension $v = (0, 0, v_0)$. For the case of a quadrupole field generated by a pair of coils as illustrated in Fig.(5.4), the equations of motion lead to a coupled system of dierential equations which can easily be solved numerically, as just the simple cases (e.g. $B = (0, 0, B_z)$) allow for analytical solutions.

$$\frac{\partial}{\partial t}v_x = \frac{q}{m} \ (v_y B_z - v_z B_x) \qquad (5.11)$$

$$\frac{\partial}{\partial t}v_y = \frac{q}{m} \ (v_z B_x - v_x B_y) \qquad (5.12)$$

$$\frac{\partial}{\partial t}v_z = \frac{q}{m} \ (v_x B_y - v_y B_z) \qquad (5.13)$$

5.3. Laser cooling and magnetic trapping of atoms

Atoms experience strong forces in light fields which can be used for radiation pressure trapping and cooling as first proposed by *Hänsch* and *Schawlow* in 1975 [67]. Laser cooling techniques which have raised up the last decades [68, 69, 70] therefore have turned trapping and cooling

---

[3]both from Scientific Industry Services Inc., Ringoes, NJ, United States
[4]Comsol Multiphysics Simulation, Burlington, MA USA

of atoms into a fundamental field of research in the area of atomic- and molecular physics.

Beginning with the first proposals [71] for magnetic trapping of neutral atoms in the 1960s, first successful trapping was reported in [72]. Both techniques, laser cooling and magnetic trapping, with subsequent cooling techniques in the magnetic traps [73, 74] opened the road towards Bose-Einstein condensation [41, 42] of cold, trapped neutral atoms in the following decade.

The following section briefly reviews the most important experimental and theoretical foundations for the presented work. For deeper insights, the reader is pointed to textbooks [75], and exhaustive reviews [76].

5.3.1. The linear Zeeman-shift

The Zeeman-eect [77], and the discrete splitting of an Ag-beam in the famous *Stern-Gerlach* experiment [78] is based on the quantized, linear level-splitting that occurs in the presence of a weak magnetic field. Magnetic trapping is therefore based on the level-splitting of atoms in magnetic fields, which allows to induce spatially dependent energy shifts. While this does not apply to zero-nuclear spin particles (even with a large electronic spin, such as ^{52}Cr), which are trapped in fine-structure states [79], non-zero nuclear spin atoms such as rubidium are trapped in the hyperfine states, facing the above mentioned Zeeman-shift.

The energy-shift $\Delta E = -\mu_B \; B$ for ^{87}Rb in a weak magnetic field which does not decouple the nuclear and electronic spins, therefore reads as

$$\Delta E = -g_F m_F \mu_B |B| \tag{5.14}$$

resulting in a force acting on the neutral atoms in a non-constant B-field

$$F = g_F m_F \mu_B \nabla B \tag{5.15}$$

with the Bohr-magneton $\mu_B = e\hbar/2m_e$, the quantum number m_F, the Lande-g factor g_F, and the magnetic field B. For ^{87}Rb atoms in the $5^2S_{1/2}$ ground-state, the trappable lowfield-seeking states for which $m_F g_F > 0$ is valid, are the $|F, m_F\rangle = |2, 2\rangle, |2, 1\rangle \; and \; |1, -1\rangle$ states, where atoms are attracted to minimum magnetic fields. Regarding to *Earnshaw's theorem* [80], configurations attracting the lowfield-seeking states are the one and only stable static magnetic traps, as Maxwell's equations do not allow for magnetic field-maxima inside a source-free volume.

5.3.2. Magneto-optical trap and optical molasses

Employing optical and magnetic fields, a MOT [81] is the working horse aiming for cold trapped ensembles of alkali-metals. Consisting of three counter-propagating circular polarized laser beams with an overlaid linearly inhomogeneous magnetic field the setup is rather simple and robust once a MOT is performed.

Based on the scattering rate of a two-level system in a near resonant laser field, deduced from the optical *Bloch-equations* [82], the scattering force on the system, can be written as

5. Theory

$$F_{scatt} = \hbar k \frac{\Gamma}{2} \times \frac{s_0}{1 + s_0 + 4\delta^2/\Gamma^2} \quad (5.16)$$

with the spontaneous emission rate Γ and the normalized intensity $s_0 = I/I_{sat}$. The laser detuning δ from the resonance frequency ω_0 is given by $\delta = \omega - \omega_0 + kv$ regarding the *Doppler-shift* kv which experiences the atomic resonance frequency at velocity v.

Caused by the linear increasing magnetic field in a MOT, the hyperfine levels are split and *Zeeman-shifted*, Eq.(5.14). Thus adding a position dependent frequency shift in the detuning, leads for the force on an atom in the MOT in one dimension to

$$F_{MOT} = F_{scatt}^+[\delta - kv - (\omega_0 + \mu z)] - F_{scatt}^-[\delta + kv - (\omega_0 - \mu z)] \quad (5.17)$$

considering two counter propagating circular polarized light beams. Neglecting the velocity term in Eq.(5.17) ($kv \ll \Gamma, kv \ll \mu$), the conservative restoring force in a MOT is proportional to the position $r = (r_x, r_y, r_z)$ with $F \propto D_{x,y,z} \cdot r$ and causes trapping of the atoms. Due to the scattering force, this leads to a velocity damping in position-space, described by the diffusion constant $D_{x,y,z}$.

The capture process of fast atoms far away from the trapping center is position and velocity dependent, but is still governed by Eq.(5.16). The atomic trajectories and therefore the capture process can easily be simulated, but follows the arguments by *Metcalf* and *Van der Straten* [83, 84] to be estimated.

Optical molasses

In comparison to the spatially dependent net force Eq.(5.17), the sum of the scattering force Eq.(5.16) of two red detuned, counter propagating laser beams in absence of magnetic fields, leads to diffusion of the atom cloud in the optical molasses [85] using circular polarized beams, and reads in one dimension as

$$F_{mol} = F_{scatt}^+(\delta - \omega_0 - kv) - F_{scatt}^-(\delta - \omega_0 + kv) \quad (5.18)$$

This process happens in velocity-space as atoms do not experience a restoring, trapping force. For atoms already cooled below a certain velocity with ($kv \ll \Gamma$) the net force can be described by a damping process instead of trapping, with $F_{net} \propto -\beta v$ in one dimension. Derived from Eq.(5.16) and Eq.(5.18), the damping constant β turns out to be

$$\beta = -8\hbar k^2 s_0 \frac{\delta/\Gamma}{(1 + 4\delta^2/\Gamma^2)^2} \quad (5.19)$$

and hence describes a friction process regarding to $m\frac{d}{dt}v = -\beta v$ proceeding in the optical molasses at a red-detuning $\delta \gg \Gamma/2$. Hence expressing the net-force in loss of kinetic energy of the atoms, this leads to cooling as the kinetic energy is reduced described by

$$\frac{d}{dt}E_{kin} = -\beta v^2 \quad (5.20)$$

Considering these trapping- and friction-forces, the Doppler-temperature $T_D = 143\mu K$ at a maximum red-detuning of $\delta = -\Gamma/2$ was thought to be the limit for Doppler cooling [86].

Nevertheless, the atoms are not simple two-level systems, and considering polarization gradient cooling [87, 88] the achievable temperatures, almost one order of magnitude below T_D, and slightly above $T_{recoil} \propto \hbar^2 k^2/2mc^2$ can be explained.

Loading rate and losses in a MOT

Loading a MOT follows an exponential law described by the rate-equation

$$\frac{dN}{dt} = L - N\gamma - \beta \int n^2(r,t)d^3r \quad (5.21)$$

following [89], with capture rate L and the loss rates for background- and internal collisions γ and β. According to [90] dominant radiation trapping at higher trap-densities causes a constant density distribution n_c, rather than a Gaussian distribution as described by $n(r,t) = n_0(t)e^{-r^2/w_{FWHM}^2}$. Therefore Eq.(5.21) turns into

$$\frac{dN}{dt} = L - R\, N \quad (5.22)$$

with the loss rate $R = (\gamma + n_c\beta) = 1/\tau_L$ and the loading rate $L = N_\infty/\tau_L$. This euqtion is covered by the solution $N(t) = N_\infty\, e^{-t/\tau_L}$. For low MOT-densities, the solution follows a more complicated way based on

$$\frac{dN}{dt} = L - N\gamma - \beta\frac{N^2}{(2\pi)^{(3/2)}w^3} \quad (5.23)$$

For an experiment where the lifetime in the MOT is measured, for a switched of atom-source and therefore $L = 0$, the solution follows [91] and can be written as

$$N(t') = N_0 e^{-t/\tau}\frac{1}{1+\beta_0 N_0[1-e^{-t/\tau}]} \quad (5.24)$$

for $t' = t/\tau = t$ and $\beta_0 = \frac{\beta}{(2\pi)^{(3/2)}w^3}$.

5.3.3. Magnetic traps for neutral atoms

Quadrupole trap

Based on the *Zeeman-e ect*, the most simple setups of a trap which possesses a magnetic moment in a static magnetic field, addresses the quadrupole trap [72] with a pair of counter-operated coils, and a simple 2D-wire trap with a homogeneous B-field superimposed orthogonal to a current guiding wire [92].
Superimposing the magnetic fields of two circular coils in a distance 2d with radius R, the magnetic field reads as

$$B(x,y,z) = A \times e_A \quad with: A = \frac{3\mu_0 I d R^2}{2(d^2+R^2)^{5/2}} \quad (5.25)$$

where $e_A = (x, 2y, z)$, providing the gradient $\partial B_y/\partial B_{x,z} = 2$ for $2d = R$.
Both traps provide a zero-trap bottom[5], which is not desirable for cooling the atoms further down using evaporative techniques [42], as Majorana-spin-flips can occur.

---

[5] $|\vec{B}| = 0$ at the trap center

5. Theory

Ioffe-Pritchard type traps

To perform a trapping potential which allows a non-zero trap bottom, two prominent realizations are possible. Either a so called time-orbiting-potential (TOP) trap [93], or a *Ioffe-Pritchard* trap. This trap was first proposed by *Pritchard* [73] for magnetic trapping of neutral atoms, while the Russian physicist *Ioffe*, working in the field of plasma physics, used a superimposed homogeneous field to maintain a non-zero field-confinement [94]. The first trapping in a *Ioffe-Pritchard* like trap was shown some years later [95, 96], and since than, dierent ways of implementing *Ioffe-Pritchard* traps have been found, covering the Ioe-type trap with permanent magnets [97], the clover-leaf [98, 99], the baseball-trap [100], the Biquic-trap [101] and others more.

All those non-zero trap bottom configurations have in common that the strength of the trapping can not be expressed in terms of the gradient (∇B), which is proportional to the trapping force Eq.(5.15) in a simple quadrupole trap.

On one hand, the Ioe-trap provides a steep confinement for atoms with $k_B T > B_0$ and for atoms close to the trap center, and even cold atoms with $k_B T \approx B_0$ a harmonic potential with $|B_{min}| \neq 0$. Hence this oset field $|B_{min}|$, defines the trap bottom B_0. In a quasi-harmonic approximation the trap frequencies $= 2f$, regarding to the frequencies of a harmonic oscillator can be defined, following the curvature along one axis in the magnetic field and reads as

$$\omega_r = \frac{2\mu_B m_F g_F}{m_{Rb}} \times \sqrt{\frac{\partial^2 B_r}{\partial r}} \tag{5.26}$$

Another Ioe-Pritchard trap should be highlighted, which is used in this thesis. This quadrupole-Ioe-configuration (QUIC) trap, is one of the easiest realizations, by applying an additional coil to a pair of quadrupole coils.

QUIC-trap

The intrinsic properties of the QUIC-trap are presented in the following. For a simple setup the transformation of a quadrupole trap into a QUIC-trap is analytically calculated and plotted, regarding the scheme illustrated in Fig.(5.5). Therefore the magnetic field around the trapping center follows

$$B(\rho, z) = B_0 + \frac{1}{2}(\frac{\rho^2}{B_0} - \frac{\rho^2}{2})^2 + \frac{1}{2}z^2 \tag{5.27}$$

with the field gradient in radial direction and the curvature in axial direction.

A trap with non-zero trap bottom B_0 can be achieved, superposing an inhomogeneous field to an existing quadrupole trap. When the QUIC-coil is o, the trap center is in the middle of the vertical quadrupole coils and the trapping potential refers to the linear quadrupole trap gradients as shown in inset a) of Fig.(5.6), and following $|B(x,y,z)| = -\frac{B'}{2}x + B'y - \frac{B'}{2}z$, respectively Eq.(5.25). If the QUIC-coil is slowly turned on, the magnetic field amplitude in Ioe-direction (z-direction) is still at certain positions below the linear gradient in radial direction, causing two field crossings, and therefore two trap-minima, both at zero field amplitude. Hence both field-minima are shifted towards the QUIC-coil as shown in inset b). At the point when the axial field component of the QUIC-coil exactly cancels the quadrupole field in Ioe-direction, the two field-minima merges and form a more harmonic trapping potential

5.3. Laser cooling and magnetic trapping of atoms

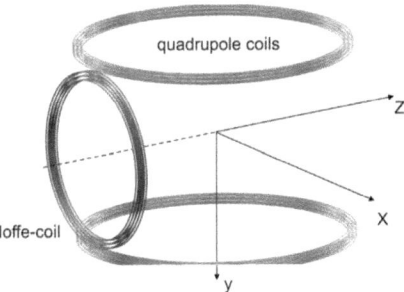

Figure 5.5.: Three coils built up from a series of single coils are superimposed to form a QUIC-trap, as shown in Fig.(5.6), ramping up a current in the Ioffe-coil.

than a linear quadrupole gradient, with the minimum still at zero. From now on the trap confinement is described by the harmonic trap frequencies rather than by a field gradient, inset c). If the QUIC-current is further increased, the trap bottom is further lifted from zero magnetic field, and slightly shifted towards the vertical quadrupole coil axis. The trap bottom B_0 is now defined by the difference in magnetic field of the two Ioffe-directed field components, inset d). The established QUIC-trap therefore provides a harmonic potential and a non-zero trap bottom, sensitively dependent on the stability of the two different coil currents in the QUIC-coil and the quadrupole-coil-pair and follows [102] with the trapping potential described by the harmonic trap frequencies $\omega_r \gg \omega_z$

$$V(\rho, z) = V_0 + \frac{1}{2}m\omega_r^2 \rho^2 + \frac{1}{2}m\omega_z^2 z^2 \quad \text{with:} \quad \rho = \sqrt{x^2 + y^2} \quad (5.28)$$

5.3.4. Superconducting micro-traps

Even if the first superconducting trap for cold atoms [40] was already demonstrated in 1995 using a macroscopic quadrupole trap with zero-trap bottom, experiments with superconducting traps even ascent since the superconducting atom chip was performed. Based on the last decades advance, exhausting work have been published [6, 7, 8, 9] considering normal conducting micro-traps and was extended by studies about superconducting micro-traps in recent years [103, 104, 105].

As shown in chapter (15), a superconducting micro-trap is used, existent of a Z-shaped wire, and a homogeneous bias field B_b. The calculation in section (15.1) is based on the far-field of the trap, which even does not account for superconducting effects, and finite size effects. Hence the properties of the superconducting micro-trap properties far away from the wire without an additional Ioffe-field B_I can simply be described, following [106] and [107]:

With the length L of the central Z-wire, and the current I_Z, the vertical distance d_m of the trapping minimum from the chip-surface is defined by

$$d_m^2 = L\frac{\sqrt{(LB_b)^2 + (2\mu_0 I_Z)^2}}{8B_b} - \frac{L^2}{8} \quad (5.29)$$

5. Theory

while the trap bottom B_0 scales as

$$B_0 \propto \frac{I_z d_m}{4d_m^2 + L^2} \quad (5.30)$$

and the curvatures in radial and longitudinal y-direction goes as

$$\frac{\partial^2}{\partial r^2} B \propto \frac{I_z(4d_m^2 + L^2)}{d_m^5} \quad (5.31)$$

$$\frac{\partial^2}{\partial y^2} B \propto \frac{I_z}{d_m L^2} \quad (5.32)$$

These proportional behavior clearly exhibit that the trap frequencies can dramatically increased for bringing the trap minimum close to the central wire, reaching transversal trap frequencies well above $\omega_\perp > 2\pi \times 1 kHz$

5.4. Interactions, collisions and loss-rates in magnetic traps for cold atoms

Majorana spin-flips

Keeping atoms magnetically trapped in a static B-field configuration, the atoms must be in the lowfield-seeking state with $m_F g_F > 0$. Therefore spin-flip transitions between Zeeman-levels due to the adiabatic approximation shown in section (5.3.1) can cause the atoms to flip into a high-field seeking state and therefore to be lost from the trap. These transitions were first studied in 1932 [108], in a one-dimensional time-dependent model by *Majorana*, and are therefore also known as Majorana-transitions, or Majorana-spin-flips. The adiabatic approximation, hence is only valid as long as the spin of the moving atom can follow the direction of the field, which defines the quantization axis. As the spin precesses around the quantization axis with the Larmor-frequency $\omega_L = |U|/\hbar$, the change in magnetic field has to be slower than that. In this case the eigenvalues of the magnetic quantum number F can be treated spatially independent, even if the quantization axis change directions in space.

$$|B|^{-1} \frac{d}{dt}|B| < \omega_L = \frac{m_F g_F \mu_B |B|}{\hbar} \quad (5.33)$$

The above given inequality is therefore the condition that atoms do not undergo non-adiabatic spin-flips which results in highfield-seeking states, loosing the atoms from the trap. For vanishing B-fields, such as in the trap center of quadrupole traps, this condition does not hold.

For Ioe-Pritchard traps with a trap bottom $B_0 \neq 0$, the non-adiabatic spin-flip ratio can be calculated regarding [109, 110]. For spin-1 particles trapped in a cigar-shaped configuration case 1), with the strong radial trapping frequency $\omega_r \gg \omega_z$ and the longitudinal trapping frequency in Ioe-direction ω_z, and case 2), atoms trapped in a spherical trap with $\omega_r = \omega_z$, the loss rate reduces to

$$M = \begin{cases} 4\pi\omega_r e^{-\frac{2\omega_L}{\omega_r}} & \text{case 1: for } \omega_L \gg \omega_r \gg \omega_z \\ 8\sqrt{2}\pi\omega_i \omega_L e^{-\frac{2\omega_L}{\omega_i}} & \text{case 2: for } \omega_L \gg \omega_i \equiv \omega_r = \omega_z \end{cases} \quad (5.34)$$

Background collisions

The magnetic quadrupole traps provide at least a trap-depth of $T_{depth} \approx \frac{\mu_B B_{max-quad}}{k_B} \approx 10mK$. Nevertheless this causes atoms to be lost from the trap if they maintain a momentum tranfer by collisions with background gas. The loss corresponds to $dN = N_0 e^{-t/\tau_{bg}}$, with the characteristic time constant τ_{bg} and the loss-rate $\Gamma_{bg} = 1/\tau_{bg}$. According to [111] this loss rate is given as

$$\Gamma_{bg} \cong n_{bg} \, \sigma_{bg} \, v_{bg} \tag{5.35}$$

for a certain background pressure p, the atom density n_{bg} in the background gas, the cross-section σ_{bg} for the dominant species[6], and the thermal mean velocity $v_{bg} = \sqrt{2kBT/m_{bg}}$ of the background gas species.

Inelastic two-body collisions

Due to the spin-spin interaction of the valence electrons or $S L_{diff}$ coupling of the spin with the relative orbital angular momentum of two colliding atoms, the magnetic quantum numbers m_F can change, releasing energy in the order of $\mu_B |B|$. As long as the atoms are not pumped into the maximum polarized state $|F=2, m_F=2\rangle$, these collisions are not suppressed and can lead to losses of atoms from the trap, regarding to the loss rate

$$\Gamma_{dip} = G_{dip} \, n_{trap} \quad \text{with:} \quad G_{dip} = 10^{-14} cm^3/s \tag{5.36}$$

which scales with the coupling constant G_{dip} and the density of atoms in the trap n_{trap} according to [112].

Inelastic three-body collisions

Even the spin-polarization prevents the building of ^{87}Rb-molecules due to two-body collisions, molecules can form if a third particle is involved which gathers the binding energy as kinetic energy-gain. The loss rate according two these three-body collisions is therefore scaling as

$$\Gamma_{3-body} = L_{3-body} \, n_{trap}^2 \quad \text{with:} \quad L_{3-body} = 4 \cdot 10^{-29} cm^6/s \tag{5.37}$$

Lifetime of trapped atoms

The three above mentioned loss mechanisms limits therefore the lifetime in magnetic traps and can be subsumed as

$$\tau_{loss} = \frac{1}{\Gamma_{bg} + \Gamma_{dip} + \Gamma_{3-body}} = \frac{1}{\Gamma_{bg} + G_{dip} \, n_{trap} + L_{3-body} \, n_{trap}^2} \tag{5.38}$$

in dependence of the atom-density n_{trap} in the trap. Fig.(5.7) illustrates the lifetime-limiting issue. At rather high pressures in the order of $10^{-10}...10^{-9} mbar$ the limiting factor are losses due to collisions with the background as the figure shows, both for H_2 or ^{87}Rb as the main dominating species in a vacuum system. For better background pressures as in

---
[6]H_2 molecules: $\sigma_{bg} = 2 \cdot 10^{-18} m^2$; ^{87}Rb: $\sigma_{bg} = 2.5 \cdot 10^{-17} m^2$ according to [111]

5. Theory

cryogenic environments usually the case, the limiting factor are the inelastic two- and three-body collisions even if they appreciable occur at densities above in the order of $10^{15}/m^3$. Fig.(5.7) therefore shows for both H_2 (dashed line) and ^{87}Rb (solid line), the lifetime regarding to Eq.(5.38) for $n_{trap} = 10^{15}...10^{16}...10^{17} \; atoms/m^3$.

A lifetime measurement therefore often exhibits a decay behavior more complex than a simple exponential decay due to the two- and three-body trap-loss collisions between trapped atoms. A deviation from $N \approx e^{-t/\tau}$ also occurs due to the loss of hot atoms which are kicked out of the trap by elastic collisions if the trap is not deep enough and hence changing the density dramatically (e.g, due to heating, or after reloading in a shallow trap). The atom-number decays with time and therefore the atom density drops down, exhibiting on longer time scales the characteristic constant for the background-losses.

Elastic collisions and thermalization

Even for the prospective evaporative cooling [42] for reloading[7] the atoms into a chip-based micro-trap to achieve Bose-Einstein condensation, even a ratio of elastic and inelastic collisions in the order of $R = \frac{\Gamma_{el}}{\Gamma_{el}} = 5$ must be at least satisfied [113, 114], as the elastic collisions are responsible for the thermalization of the atomic ensemble.

The rate of elastic collisions can therefore be described using the mean relative velocity of two atoms $\bar{v} = \sqrt{8k_B T/m_{Rb}}$, the mean atom-density $\bar{n}$ and the cross-section $\sigma_{el} = 8\pi a^2$ which for small temperatures[8] is energy-independent. According to the measured peak-density of the atom-cloud, Γ_{el} reads as

$$\Gamma_{el} = \sqrt{2} \, \bar{n} \, \sigma_{el} \, \bar{v} \begin{cases} \text{with:} & \bar{n} = \frac{n_0}{8} \quad \text{for linear trapping gradients} \\ \text{with:} & \bar{n} = \frac{n_0}{\sqrt{8}} \quad \text{for harmonic confinement} \end{cases} \quad (5.39)$$

Heating rate

Beside the heating of trapped atoms by accidentally shined in, resonant laser light, another mechanisms leads to heating, which is obstructive anyway as atoms could be kicked out of the trap.

From the model Hamiltonian

$$H = \frac{p^2}{2m_{Rb}} + \frac{1}{2}m_{Rb} \, \omega_{trap}^2 [1 + \epsilon(t)]x^2 \quad (5.40)$$

for a trapped Rb-atom, with the mean-square trap frequency ω_{trap}^2 and the time-dependent fractional fluctuations $\epsilon(t)$ in trapping currents. After treatment with first-order time-dependent perturbation theory, this simple model [116, 117] exhibits the energy e-folding time constant $\tau_{heat} = 1/\Gamma$ for the exponential growth in average particle energy which is described by $\langle \dot{E} \rangle = \Gamma \langle E \rangle$ and results in

$$\Gamma = \frac{\pi}{2} \omega_{trap}^2 S(2\omega_{trap}) = \pi^2 \nu_{trap}^2 S(2\nu_{trap}) \quad (5.41)$$

---

[7]For the reloading process see section 15.1

[8]For temperatures $T < 100\mu K$ in ^{87}Rb, just s-wave scattering occurs, where the scattering length of the triplet state is $a = 106 \cdot a_0$; [115]

28

5.4. Interactions, collisions and loss-rates in magnetic traps for cold atoms

with the trapping frequency $\omega_{trap} = 1\ Hz$, $\tau_{heat} = 1\ s$, and the power spectrum $S(\omega)$ of the fractional magnetic field noise. An estimate for the power spectrum is therefore $S(\omega) = \epsilon^2/\Delta\omega$ which is the RMS-fractional square in a bandwidth of $\Delta\omega$.
For a simple quadrupole configuration and for $\epsilon = \frac{I(t)-I_0}{I_0}$ the e-folding time can be calculated according to

$$\tau_{heat} = \omega^2 \omega_{trap}^2 \frac{\epsilon^2}{\Delta\omega} \qquad (5.42)$$

which results for $\Delta\omega = 1$kHz, and $\omega_{trap} = 100$Hz in

$$\tau_{heat} = \left[\frac{10^{-2}}{\epsilon^2}\right] s \qquad (5.43)$$

and therefore $\tau_{heat} = 100s$ if a power-supply stability of 1% is assumed. Things looks completely different if a QUIC-trap is considered, with the fractional power-supply stabilites of $\epsilon_i = \frac{I_i(t)-I_{0i}}{I_{0i}}$ for both currents, driving the Ioffe-coil and the quadrupole coil-pair. The overall trap-stability can therefore be written as

$$\epsilon_{tot} = \frac{\Delta I(t) - \Delta I_0}{\Delta I_0} \begin{cases} \text{with:}\ \Delta I_0 = I_{01} - I_{02} \\ \text{with:}\ \Delta I(t) = I_1(t) - I_2(t) \end{cases} \qquad (5.44)$$

Therefore the trap bottom is defined by the difference in nominal current ΔI_0 and the real current $\Delta I(t)$. With the fractional trap bottom current $\epsilon_{I0} = \frac{I_{01}-I_{02}}{I_{02}}$, the total fluctuations result in

$$\epsilon = \epsilon_1 + \frac{\epsilon_1 - \epsilon_2}{\epsilon_{I0}} \qquad (5.45)$$

and exhibits the overall stability for a QUIC-trap configuration with $I_1(t)$ corresponding to the Ioffe-current and $I_2(t)$, the quadrupole coil current.
For the case that the fractional power stability of both power-supplies would be completely equal, or even the case that the power-supply noise would be in phase, or the same current runs through the coils, $\epsilon \to \epsilon_1$ and the QUIC-trap would not be more unstable than a single-coil setup. Nevertheless two independent currents in the coils together with the fact, that the QUIC setup provides a trap bottom resulting from the difference in too huge currents, with $\epsilon_{I0} \ll 1$, this would cause extreme heating rates as Eq.(5.42) would result in $\tau_{heat} \approx 6.25s$ for the plausible case of $\epsilon_1 = 0.13\%$, $\epsilon_2 = 0.10\%$ and $\epsilon_{I0} = 0.7\%$.
Similar calculations for the heating in wire traps can be done regarding to [118] and [119].

5. Theory

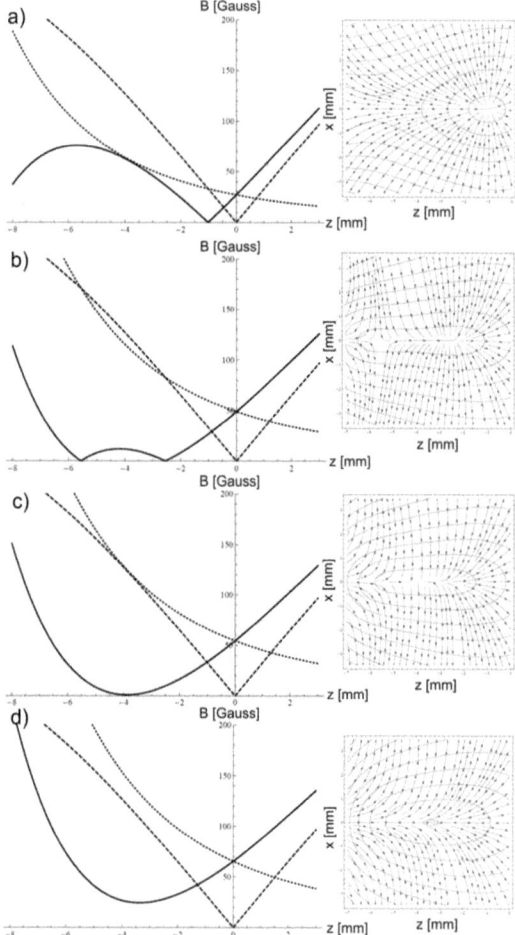

Figure 5.6.: A quadrupole-Ioe-configuration (QUIC) -trap with non-zero trap bottom can be achieved, superposing an inhomogeneous radial field to an existing quadrupole trap. Inset a)-d) gives for $I_{Ioffe} = 0$, and $I_{Ioffe} = I_{1-3}$ the magnetic trapping potential for this setup, illustrating how the QUIC-trap forms. The insets on the right, additionally show the magnetic field lines corresponding to the graphs on the left in the horizontal plane.

5.4. Interactions, collisions and loss-rates in magnetic traps for cold atoms

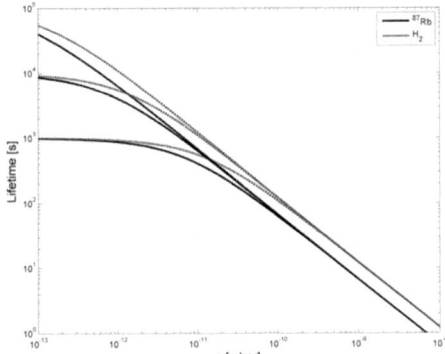

Figure 5.7.: Lifetime in dependence of the background pressure for the dominant species H_2 and ^{87}Rb for dierent trap-densities $n_{trap} = 10^{15}...10^{16}.10^{17}\ atoms/m^3$

Part I.
Electron-beam driven atom source for cryogenic environments

6 Concept of an electron-beam driven atom source

Dierent alkali atom sources for cold atom experiments beside many others [47], rely on light induced atomic desorption (LIAD) from dielectric surfaces and glass [120, 121] or stainless steel [122], or on thermal heated alkali metal dispensers (AMD's) [123, 124], laser ablation [125] or laser heating [126]. Under certain conditions, cryogenic, buer gas cooled sources could provide sodium beams as demonstrated in [127].

An *electron beam driven atom source* in this manner should be considered as a source of atoms which is, due to the interaction of an electron beam with a surface, released from the latter. As pointed out in section (4.3) such an electron beam driven atom source for alkali metals, may satisfy the demands for an atom source for cryogenic systems. This already implicitly set the demands for the atom source, which should on one hand be capable to set free a certain amount of atoms, and on the other hand just consume a minimum power, compatible with the cooling power of a cryogenic system at 4K, usually below 1W.

Electron stimulated desorption (ESD)

In principle two potential processes exist to desorb alkali metals from surfaces, by the use of electron beams. As shown in section (9.3), several experiments were performed [128, 129, 130, 131], to investigate electron stimulated desorption [132, 133, 134] (ESD) of Li, Na, K and Cs from alkali metal layers on oxidized W- or Mo- surfaces, since this eect is almost known for more than 50 years, and was once in a while re-interpretated [135, 136]. This process relies on incident electrons, which creates in the 2s oxygen level a core-hole which triggers an intra-atomic Auger-decay which leads to a subsequent neutralization of one positive alkali metal ion. If therefore the positive oxygen ion can capture electrons from the metal-substrate to achieve a negative charge state again, the alkali metal atom will be repelled and desorbed from the surface as a neutral atom. ESD is therefore in principle similar to Photon stimulated desorption [137, 138] (PSD), which is accountable for sodium in the lunar atmosphere [139].

6. Concept of an electron-beam driven atom source

Electron beam physical vapor deposition (EBPVD)

Keeping in mind the electron-bulk interaction processes discussed earlier in chapter (5.1), the electron bombardment leads to a heating of the target where a part of the electrons kinetic energy is converted into thermal heat. From electron beam physical vapor deposition (EBPVD), or simply e-beam evaporation, this heating is a well known and a used technique to industrially coat surfaces with a high evaporation rate.

It should preliminary be mentioned that the results presented in section (9) rather indicates that ESD is the relying process which causes the atoms to be desorbed from the target surfaces. Nevertheless the results are rather indications for ESD, than a strong evidence, as the setup did not allow for detailed investigations, which also was not the intention for this experiment. Nevertheless the subsequent modeling which exhibits the scaling laws and the parameter range, and a performance of an evaporation based desorption, finally allows to exclude this process, and therefore strongly supports the conclusion in chapter 16.

6.1. Motivation for modeling the desorption of atoms based on EBPVD

In electron beam physical vapor deposition, electron beams up to hundreds of kW electron power, focused to an area of some square centimeter heat up huge crucibles filled with nearly arbitrary evaporants or compounds. A full range of elements can be applied to EBPVD such as $Au, Ag, Al, Sn, Cr, Sb, Ge, In, Mg, Ga, NaCl, KCl, AgCl, CaF_2, MgF_2$ which are resistive heated, extended by insulators and oxide compounds such as Al_2O_3, SiO SiO_2, SnO_2, TiO_2, ZrO_2 and even high melting metals such as $Ni, Pt, Ir, Rh, Ti, V, Zr, W, Ta, Mo$. Thinking of the rather high vapor pressure of alkali metals, where metal-like free electrons sustain electric conduction, it is obvious that those alkali metals could be evaporated using electron beams.

In Fig.(6.1), the vapor pressure of several alkali metals is depicted, compared with those of common other evaporants[1]. As the reason for the high powers used in EBPVD are in fact the low vapor pressures at certain temperatures and the good thermal conduction of metals, it is clear that for alkali metals even a moderate heating can lead to a significant evaporation.

Therefore the following chapter enlightens the models used to describe the release of atoms from a surface due to thermal heating by a focused electron beam bombarding a target. A 2D-model [145] is used to calculate surface heating and a standard model for thermal evaporation [146] than leads to an atom-flux above the surface. Followed by the investigation of crucial parameters for this process, also limiting parameters are under the looking glass.

6.2. Electron beam impinging a surface

Due to a certain kinetic energy of the incident electron beam and therefore a resulting penetration depth, interactions of electrons with the bulk take place in a certain interaction volume. As described earlier in section (5.1), electron collisions depend on the beam energy, the incident angle as well as on the material properties. Furthermore the incident beam causes

---
[1]Vapor pressures are taken from: Rb:[140], Cs:[141], Na:[142], Cu:[143], Li:[144], Al, Ti, Mn from common literature

6.2. Electron beam impinging a surface

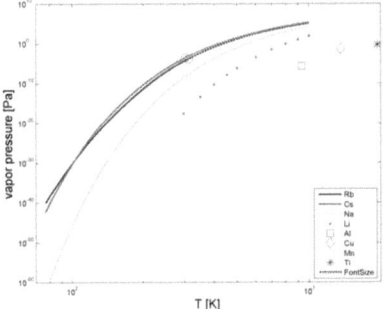

Figure 6.1.: The vapor pressure of alkali metals (Li, Na, Rb, Cs) in contrast to dierent common elements (Cu, Al, Ti, Mn) used in electron beam evaporation processes is shown. Solid lines are based on analytical functions related to Rb:[140], Cs:[141], Na:[142] and Li:[144]. In addition find data for Cu in [143], and common literature

secondary electrons, as well as electrons can either be backscattered or absorbed regarding to section (5.1), where in the following the emission of x-rays should be neglected [145]. Hence, a certain number of electrons, proportional to the net-bulk current than dissipate their energy through elastic and inelastic collisions into thermal heat.

In the following it will be assumed that just absorbed electrons lead to thermal heating of the bulk in a region defined by the penetration depth of the electrons. Even for this experiment this seems quite reasonable as this assumption would be an underestimate of the achieved heating[2].

6.2.1. Penetration depth of electrons

The collision processes mentioned above are purely random, and a calculation of the penetration depth can be performed with a Monte-Carlo Simulation[3] deriving the penetration depth from a huge number of single electron trajectories. Therefore the simulation program CASINO [147] was used to simulate a series of trajectories depicted in Fig.(6.2). Additionally the program extracts the number of backscattered electrons which is derived to be $_{back} = 0.345$. As the program evaluates the profile over all trajectories, mapping the maximum depth of each electron, the extruded penetration depth is therefore the mean of the maximum depth of each trajectory.

---

[2]So far this approximation would be an underestimate as backscattered electrons ($E > 50eV$) would also contribute to the heating. In this manner, secondary electrons and Auger-electrons would negligibly contribute to a heating, keeping in mind, that the incident electron beam is in the energy range from 1-6kV. Additional it should be noticed that even backscattered, and secondary electrons could be recycled by using a magnetic field (e.g behind the target) which bends the escaping electrons back towards the surface.

[3]Common programs like the CASINO [147] accurately simulates single electron trajectories due to the well known understanding of the mean free path of electrons in solids [50] and scattering cross sections

6. Concept of an electron-beam driven atom source

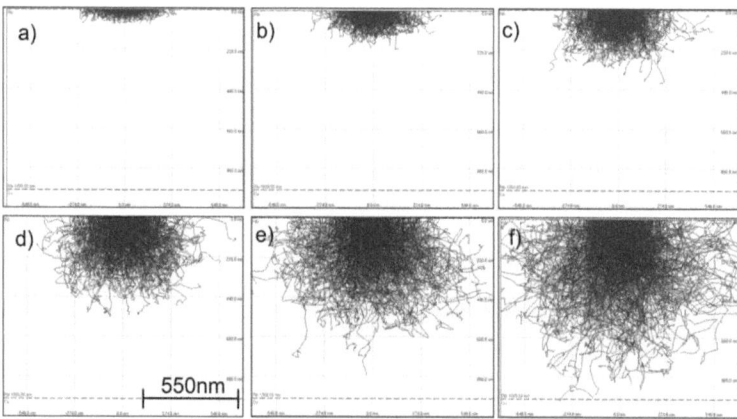

Figure 6.2.: Insets a)-f) show the simulated trajectories of electrons with a kinetic energy of 1keV, 2keV, 3keV, 4keV, 5keV and 6keV impinging a 1 μm thick Rb layer on top of an Cu bulk

CASINO underlies well tested models for calculating the total and partial cross-sections where the simulation data than refers to different used models. For completeness the models in the above described simulation are given[4]

According to the theory of electron scattering cross sections - essentially for scanning electron microscopy [148, 149], the dependence of the penetration depth with kinetic energy is well understood and can be described by a more analytical model found by *Kanaya* and *Okayama* [51]. This model gives the penetration depth as the radius of a circle located at the surface center of the beam impact on the target, estimating the interaction volume of the electrons. For the sake of completeness (see section 5.1.2) the formula is given again, reading as

$$D_{KO} = 0.0276 M \times \frac{U^{1.67}}{Z} \tag{6.1}$$

using $[M] = amu$, $[\ ] = g/cm^3$, $[U] = kV$, and $[Z] = 37$ for Rb, $[D_{KO}] = \mu m$. The dashed black line in Fig.(6.3) therefore gives the penetration depth for accelerating voltages from 1kV up to 6kV, regarding to the *Kanaya-Okayama* model. In comparison, the solid line is a quadratic fit $D_{fit} = a + bU + cU^2$ with $a = 3.87$, $b = 21.98$, $c = 8.75$ to the simulated data ($\Diamond$). To rely on a flexible model, for later considerations the quadratic fit from the Monte-Carlo simulation is used.

---

[4]For total and partial cross sections the model is *MOT by Interpolation*, for the effective section ionization *Casnati*, for the ionization potential *Joy and Luo*, and *Press et al* for the random number generator.

6.2. Electron beam impinging a surface

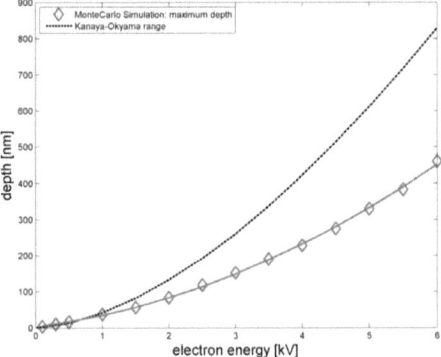

Figure 6.3.: Mean maximum penetration depth of electrons in a rubidium target for dierent acceleration voltages, regarding to a simulation based on a Monte-Carlo algorithm

6.2.2. A 2D surface model for heating a thin film

The surface is hit by a focused electron beam, where each electron has a kinetic energy and underlays an absorption e ciency $_A$, in principle known from Eq.(5.2) and Eq.(5.3) for a wide region of atomic number Z. As an absorbed electron dissipates the energy down to a certain depth, it seems appropriate to add up the input energy per unit time to calculate the input heating power.

Once noticed that the heating of a thin film can be treated as a simple problem of heat transport in a solid, the impinging electron beam can be seen as a heat source with a certain power density and e ciency $_A$, with the power density written as

$$Q = \frac{P_{heat}}{r^2} = \frac{\dot{n}eU}{r^2} \quad _A \quad (6.2)$$

where U is the kinetic potential of the electrons and r the Gaussian half width spot-size of the electron beam. Therefore $\dot{n}$ is the number of incident electrons per second in the beam and e the elementary charge.

The subsequent calculations follow a model by Lin [145], which considers the electron beam in exactly that way: kinetic electron energy as heat source in a thin target film. The underlaying considerations and assumptions are as follows:

- The electron beam spot on the target has a finite size. Hence the electron current density is axially symmetric and shows a Gaussian distribution in radial direction.

- The conversion of kinetic energy into thermal energy takes place within the thickness D of the rubidium layer. Hence no transmission of electrons through the target is allowed.

6. Concept of an electron-beam driven atom source

- Therefore the heat is homogeneous supplied over the heated volume in z-direction
- The heat loss is independent from the planar angle and the z-direction
- The heat loss from the spot size is exclusively given by heat conduction inside the target-material.
- The temperature distribution in the spot is spatially uniform and just a function over time
- the material properties are constant over the temperature, time and space

Therefore the heat-conduction equation to be solved is written as follows:

$$\rho C \frac{\partial T(t)}{\partial t} = \kappa \nabla^2 T(t) + q - u \qquad (6.3)$$

where ρ is the density of the heated film, C the specific heat capacity, κ the thermal conductivity, q the heat source and u the heat sink per unit time and volume. An analytical temperature dependence at the beam center of the spot can easily be calculated by introducing the dimensionless variables Θ and τ with

$$\Theta = 4D \frac{\kappa T}{Qr^2} \qquad \tau = \frac{4\kappa}{C\rho r^2} t \qquad (6.4)$$

Considering conduction of heat in the Rb-bulk as the only mechanism which decreases the temperature in the center and hence neglecting losses u such as thermal radiation, the total heat source density Q $[W/m^2]$ in the heated spot size is $Q = P_{heat}/r^2$ as in Eq.(6.2). Spatial dimensions are the spot size r and the vertical dimension D of the heated volume. Regarding to the ambient temperature T_0, the absolute temperature increase ΔT after t is therefore $\Delta T = \Theta = T - T_0$, and can be written as

$$\Theta(\tau) \approx ln(1+\tau) \qquad (6.5)$$

which is the solution of Eq.(6.3).

At this point it should not be forgotten that if a material is heated from the solid phase up to the liquid phase, the heat of fusion must be considered. As for rubidium the melting point is very low ($T_M = 39.31°$) and the vapor pressure at this temperature is about $2.2610^{-6} mbar$ [140], the heat of fusion H_F can be considered as a heat sink if $T(t)$ becomes larger than the melting temperature. This can easily be accounted for in the model derived above, introducing the heat of fusion as an additional oset temperature H_F/C. Therefore the dimensionless temperature in Eq.(6.4) is extended to

$$\Theta(\tau) = \frac{4D}{Qr^2}(\kappa T + \frac{H_F}{C}) \approx ln(1+\tau) \quad \text{if: } T > T_M \qquad (6.6)$$

With Eq.(6.6) and Eq.(6.4) and inserting the power density from Eq.(6.2) the absolute temperature of the center of the spot size then reads as

$$T(t) \approx \frac{I_T U}{4\kappa D} ln(1 + \frac{4\kappa t}{C\rho r^2}) + T_0 - \frac{H_F}{C} \qquad (6.7)$$

6.2. Electron beam impinging a surface

With regarding to chapter (5), where η_A describes the absorption efficiency of electrons by the target, the relation $\eta_A \approx \frac{I_T}{I_{EB}}$ between the target current I_T and the electron beam current $I_{EB} = \dot{n} \cdot e$ is no longer important, since the target current I_T is a) a measurable quantity and b) proportional to the number of electrons absorbed in the target.

The spacial dimension D now turns out for practical reasons to be the penetration depth D of the electrons determined in section (6.2).

Hence, this allows to study the heating of a thin film varying the focused spot size at various input powers. Fig.(6.4) shows the generated spot-size temperature in dependence of the electron beam half-width diameter for several electron beam powers (black solid line). It is obvious that a smaller spot size leads to a higher heating since $T \approx P_{heat} \times const \times (ln(1/r^2))$ as well as for higher input. The set of curves in Fig.(6.4) is given for input powers reaching in steps of 10mW from initial 10mW to 60mW.

The graph also shows, that an important considerations is fulfilled even for high offset temperatures. As with decreasing spot-size the temperature rises, the radiation losses regarding Stefan-Boltzmann increases with $P_{rad} \propto T^4$. Therefore it should be noticed that neglecting the radiation losses for deriving the solution for Eq.(6.3), is even a good assumption. Fig.(6.4) shows that this is valid for a target at 4K (red dashed line) as well as for a target at $T_0 = 300K$.

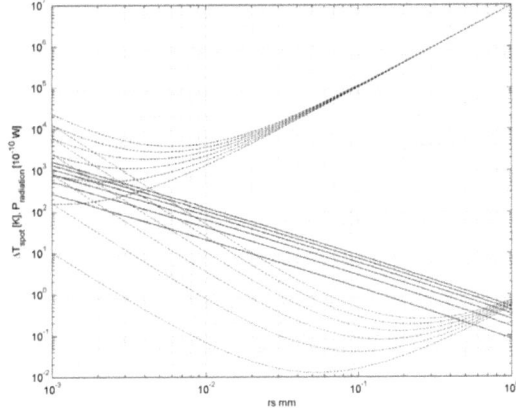

Figure 6.4.: Temperature of the heated spot after $t = 1s$ in dependence of the focused RMS beam size r_B. Curves (-black) are given for different input powers (10-60mW). For hot and very small sized spots, as well as for huge spots, the radiation losses (–) with 4K target temperature (red) and 300K target temperature (blue) the radiation losses can without restrictions be neglected.

The heating itself is a rather fast process. The temperature after 100ms can at least be seen as final temperature on a time scale of seconds, defined /C . Fig.(6.5) therefore shows

6. Concept of an electron-beam driven atom source

for a constant input power of 50mW the relative temperature increase with time for dierent RMS spot sizes of the impinging beam.

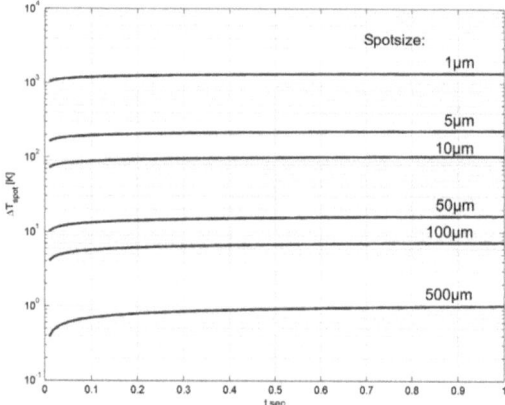

Figure 6.5.: Relative temperature increase of the heated spot size with time for an input power of 50mW.

6.3. Evaporation model and crucial parameter

To calculate the rate of evaporation after the electron beam has heated up the target material, a proper model must be found. According to Heavens [150] and [146], the rate of evaporation G $[kg/m^2s]$ of a molten drop of metal can be found as

$$G = p\sqrt{\frac{M}{2RT}} \qquad (6.8)$$

where M is the molar mass $[kg/mol]$, T the absolute temperature of the droplet, p the vapor pressure and R the gas constant in Eq.(6.8) also known as the *Langmuir-law*. This relation was tested in the mid 1950s, and even proved for metals like Fe, Ni, Mo, Ag, Ta, W and Pt as shown in [150]. Within this study, the mass-flow was measured following Eq.(6.9) and proving that C just depends on the geometry. As this is the case except for Cd and Zn, I will also use this for rubidium, without major restrictions assumed.
As we are interested in the flux of atoms penetrating an area A in the inter-crossing region of the MOT-beams, at the distance d from the target surface, the evaporated mass flow $\dot{m}$ $[kg/s]$ at the region of interest results in

6.3. Evaporation model and crucial parameter

$$\dot{m}_{evap} = Cr^2 \frac{G}{d^2} \qquad (6.9)$$

if the evaporation takes place from a sphere of radius r, with the restriction $r \ll d$.
Up to this point, we have not identified the evaporation constant C, respectively C' which is not a trivial task.
Assuming the given geometry ($d = 5.5 cm$) in the experiments [150], C' reads as $C' = C \frac{1}{d^2 2R}$ and Eq.(6.9) turns into

$$\dot{m} = C'r^2 p \sqrt{\frac{M}{T}} \qquad (6.10)$$

As in principle Heavens [150] assumes that C is constant or varies within less than one order of magnitude, one can as well take an approximate value of C to calculate the mass rates of other metals/alkaline metals, if the vapor pressure is known.

Therefore the number of atoms arriving per unit time and area A in distance d can be calculated using

$$\dot{m} = C''r^2 p \sqrt{\frac{M}{T}} \qquad (6.11)$$

with a new constant $C'' = C' \frac{d'^2 A}{A' d^2}$ where C' is now taken from literature [150] and equals 1.8×10^{-5}. With the relation $\dot{N}_{Rb} = \frac{\dot{m}}{M_{Rb}} N_{Avo}$, Eq.(6.11) turns into

$$\dot{N} = \frac{C''}{M_{Rb}} N_{Avo} r^2 p \sqrt{\frac{M}{T}} \qquad (6.12)$$

This leads effectively to an evaporated number of atoms per unit time $\dot{N}$ that reaches the MOT region. In fact the size of N is mainly affected by the vapor pressure [140] which covers many orders of magnitude in dependence of the temperature. Figure 6.6 therefore shows, the flux at the MOT region from a spot size with $r = 1\mu m$, $r = 10\mu m$, $r = 100\mu m$, for different input powers.

Additionally a correction in Eq.(6.12) must be introduced taking into account that the evaporation from a droplet is not an isotropic process [151].
According to *Knudsen* [152], the flux of an evaporation source would be described like $f(\theta)d\Omega = \cos(\theta)d\Omega$, where $f(\theta)d\Omega$ is the relative amount of material emitted within the solid angle $d\Omega$ in direction θ. As for electron beam evaporation the angular characteristics looks different, one can rely on the equation of *Graper* [153] where the evaporation can be divided into an isotropic and an non-isotropic part described by

$$f(\theta)d\Omega = [(1-A)\cos^n \theta + A]d\Omega \qquad (6.13)$$

where A is an empirical factor for the isotropic component. This factor is rather small ($A \approx 0.1$) [153] and represents the virtual source which establishes above the melting droplet. Depending on the rate of evaporation and input power, n is between 2...6 where for our case n is chosen 2.
This leads to an angular distribution of the desorbed mass flow. Integrating over the aperture, either given by the MOT trapping region and the distance of the MOT from the target spot,

6. Concept of an electron-beam driven atom source

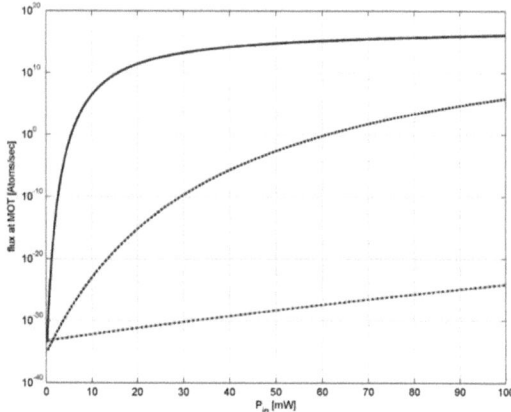

Figure 6.6.: Atom flux at the 1-inch MOT center at a distance (d=60mm) from the evaporation spot size, covering a spot size radius of $r = 1\mu m$, $r = 10\mu m$, $r = 100\mu m$ after a heating time of 1sec.

or introduced by baffles which restricts the desorbed mass flow, yields in the fractions of atoms reaching the area of the MOT. This factor $_{geom}$ can be written as

$$_{geom} = \frac{1}{2} \int f(\ )d \qquad (6.14)$$

Therefore combining the model for heating of a thin film (Eq.6.7) with the evaporation model (Eq.6.12), the dependence of atom flux with beam power, and spot size and therefore the effective temperature can be studied. The formula so far reads as

$$\dot{N}_{Rb}(t,r,T) = \ _{geom}C'\frac{d'^2 A}{A'd^2} \times \frac{N_{avo}}{M_{Rb}}r^2 p(T)\sqrt{\frac{M}{T}} \qquad (6.15)$$

and with Eq.(6.7) this turns into

$$\dot{N}_{Rb}(t,r,T) = \ _{geom}C'\sqrt{M}\frac{d'^2 A}{A'd^2} \times \frac{N_{avo}}{M_{Rb}}r^2 p(T) \times \left[\frac{I_T U}{4D}\ln(1+\frac{4t}{Cr^2})+T_0-\frac{H_F}{C}\right] \qquad (6.16)$$

where the non-isotropic evaporation behavior represented by $_{geom}$ is considered. In equation (6.15) C', d' and A' are constants which are taken from [150]. Nevertheless before the atom flux can be investigated, another inhomogeneous behavior has to be considered.

6.3. Evaporation model and crucial parameter

6.3.1. Simulation of the effective spot size

Assumptions as they where made in subsection (6.2.2), rely on a Gaussian distributed electron beam with a spotsize r_{spot} over which the heating is spatially homogeneous.
Nevertheless heat conduction along the target also leads to a temperature raise in a wider region than defined by r_{spot}. Figure 6.7 shows a finite element method simulation done with FEMLAB[5] that calculates a 2D-rotational symmetric temperature distribution over a Cu target with a $2\mu m$ thick layer of Rb on top. It clearly shows, that the assumption of a spatially homogeneous temperature-distribution in the spot size is valid as shown in inset a) which gives the surface temperature decreasing with distance. It also shows that even the surrounding region accounts to the evaporation of Rb.

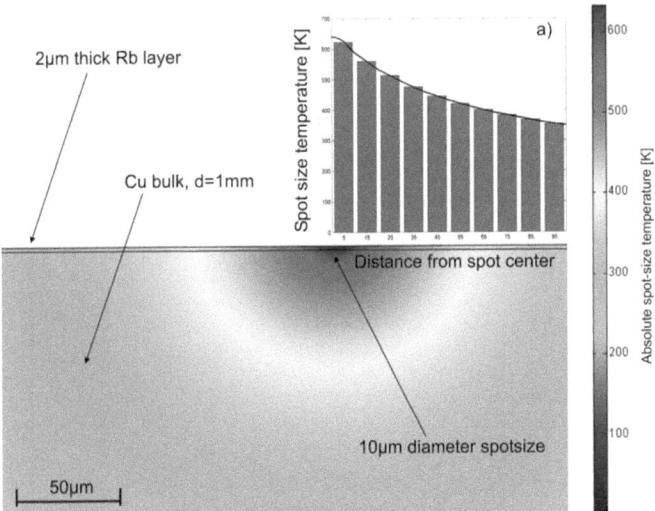

Figure 6.7.: 2D-rotational symmetric temperature simulation of the spot-size and the Cu-bulk using a finite element method (FEMLAB). Inset a) shows how the mean temperature of extended regions can be accounted for in the further calculations.

Regarding the partial pressure of rubidium, atoms are also desorbed from outside the previous defined spot size, from a circular ring shaped area dA with a spatial dependent temperature $T(r)$ shown in inset a), figure 6.7. This can be accounted for if the atom flux from the spot size is expanded for $T \to T(r)$ and therefore an integration of the gradient over

---
[5]The simulation calculates the 2D rotational-symmetric temperature distribution of a Cu bulk with a $10\mu m$ wide region of input power (200mW). It shows a consistent spot size temperature, comparable to values shown in the above given sections. The thermal bath of the Cu-bulk($2 \times 2 \times 2mm$) is set to 300K.

45

6. Concept of an electron-beam driven atom source

a wider region, written as

$$\dot{N}_{total}(t,T) = \dot{N}_{Rb}[t, r_{spot}^2, T\ ] = \int_0^\infty \frac{\partial \dot{N}_{Rb}[t,r,T(r)]}{\partial r} dr \qquad (6.17)$$

or if the dierent regions are discretized and summed up like

$$\dot{N}_{total}(t,T) \approx \sum_{i=1}^n \dot{N}(r_i, t, \bar{T}_i) \qquad (6.18)$$

where as in Eq.(6.18) this leads to a circular ring shaped area of evaporation for $i > 1$ as depicted in Fig.(6.17), inset a). This shows how the evaporation from a wider region can be taken into account building the mean of discretized *equi-thermal* ring-shapes.

Considering the loading of atoms into a MOT, only atoms with a thermal velocity below a certain capture velocity can be trapped and cooled by the magnetic confinement and the laser light as pointed out in section (5.3). As the atoms are evaporated by a surface with a temperature T this raises the question of the velocity distribution of the released atoms.

6.3.2. Temperature distribution of atoms

The trapping velocity of a MOT is given by $v_C \approx 2\frac{\Gamma}{k}(6\frac{\Gamma}{k})$ [84]. This would yield for rubidium in a capture velocity $v_C \approx 10m/s$ ($30m/s$) which has to be compared to the mean velocity of the evaporated rubidium atoms.
Described by the *Maxwell-Boltzmann* distribution in 3D, the velocity distribution can be written as

$$F(v) = \sqrt{\frac{2}{\pi}} \left(\frac{m_M}{k_B T}\right)^{3/2} v^2 e^{-\frac{m_M v^2}{2k_B T}} \qquad (6.19)$$

considering a thermal gas of rubidium atoms. As the velocity distribution is normalized to *1*, the fraction of atoms below a certain capture velocity v_C can then be written as

$$f_{\leq v_c} = \int_0^{v_c} F(v) dv \qquad (6.20)$$

Nevertheless this means, that for high temperatures $T_{mean} >> T_{capture}$ only a small fraction of the released atoms is loaded into the MOT, whereas the other atoms are lost. To account for the calculation of the trappable number of atoms $\dot{N}_{tot} \rightarrow \dot{N}_{trap}$, the trappable fraction $trap(T) = f_{\leq v_c}$ is defined. From now on, the spatial gradient of the temperature enters in the calculations. This consideration would lead us for the estimation of the total trappable atomflux[6] to

$$\dot{N}_{trap} = \int_0^\infty \frac{\partial}{\partial r} \left\{\ trap[T(r)]\dot{N}_{Rb}[t,r,T(r)]\right\} dr \qquad (6.21)$$

---

[6]Introducing the *total trapable atomflux* $\dot{N}_{trap}$, one should notice that just ^{87}Rb is trapped in the MOT which has a natural abundance of 27.8%

6.3. Evaporation model and crucial parameter

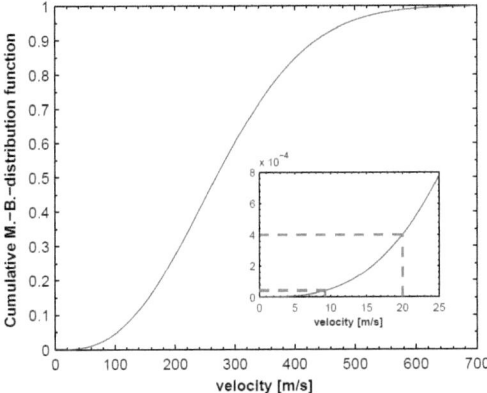

Figure 6.8.: Fraction of trappable Rb atoms. A plot of the integral over a Maxwell-Boltzmann distribution is shown. Assuming a capture velocity of $9.36 m/s$ and $(20 m/s)$ only a fraction of $4E-5(4E-4)$ atoms shown in the inset is within the capture range at T=300K. The figure is taken from [154]

where ϵ_{trap} equals the fraction of trappable atoms from Eq.(6.20) and is depicted in Fig.(6.8). As the temperature gradient $\frac{\partial T}{\partial r} = f(r)$ is a function of space, Eq.(6.17) would turn into a more complicated integral, written as

$$\dot{N}_{trap}(t,T) = \int_0^\infty \frac{\partial}{\partial T} \left\{ f(r) \; \epsilon_{trap}(T) \; \dot{N}_{Rb}(t,r,T) \right\} dr \qquad (6.22)$$

This is simply done to extend the spot-size and to account for and implement the fitted temperature gradient $T'(r) = f(r)$. To avoid such complicated derivation, where $\dot{N}_{tot} \propto \partial/\partial T [e^{-\frac{a}{T}} p(T) \frac{1}{T}]$, and subsequent integration in dr, it looks more applicable if the temperature over the spot size is discretized as described above, where the number of trappable rubidium atoms is summed up as shown in Eq.(6.18). This will make the calculation of $\dot{N}_{trap}$ easier[7], as it therefore can be written as

$$N_{trap}(t,T) \approx \sum_{i=1}^n N_{Rb}(r_i, t, \bar{T}_i) \; \epsilon_{trap\,i}(T_i) \qquad (6.23)$$

the discretized sum over di erent circular rings where the fraction enters in every partial sum as ϵ_{trap-i}.

---

[7] The integral over various variables and their derivation depending on r and T than turns into a simple sum (¡10) over the di erent ring shaped contributions.

6. Concept of an electron-beam driven atom source

6.3.3. Evaporation of particles - Mass flow calculation

Considering all influences described in the sections above, therefore leads to

$$\dot{N}_{trap}(t,T) \approx \sum_{i=1}^{n} \dot{N}_{Rb}(r_i, t, \bar{T}_i) \,_{trap\,i}(T_i) \tag{6.24}$$

Starting with the first circular shaped area where $i = 1$, the sum can be written as

$$\dot{N}_{tot}(T) \approx \,_{trap\,1}(T_1)\dot{N}_{Rb}(r_1, \bar{T}_1) + \,_{trap\,2}(T_2)\dot{N}_{Rb}(r_2, \bar{T}_2) + ... \tag{6.25}$$

if the desorption is considered for a fixed time t, and hence develops to

$$\dot{N}_{tot}(T_i) \approx \,_{i=1}^{n} c_{i1} \frac{p(\bar{T}_1)}{\sqrt{\bar{T}_1}} \left(\frac{1}{\bar{T}_1}\right)^{3/2} e^{-\frac{c_{i,2}}{T_1}} \tag{6.26}$$

The proportionality of $\dot{N}_{tot}(T)$, is therefore of interest. Given by

$$\dot{N}_{tot,i}(r_i, \bar{T}_i) \propto c_{1,i} \;\; r_i^2 \frac{p(\bar{T}_i)}{T^2} e^{-\frac{c_{2,i}}{T_i}} \tag{6.27}$$

the input power, and therefore the electron current at a certain kinetic energy turns out to be the most influencing parameter as it strongly influences the temperature at a certain spot size.

The above given discussion finally leads to the calculation of the number of trappable atoms in the MOT. For the experiment, the beam current and the spot size are crucial, and a study is illustrated Fig.(6.9). Crucial therefore means, that the parameter has to be chosen in the right way to evaporate a feasible amount of atoms which can be trapped in a MOT. On the other side there may be a limiting parameter defined, which is the surrounding target temperature.

Fig.(6.9) shows how the atom flux varies over more than three orders of magnitude if the electron current is increased from $2\mu A$ up to $10\mu A$. As at a constant input power the temperature decreases with larger evaporation spot size, also the fraction of trappable atoms gets bigger, whereas at small spot sizes a high temperature does not exclusively lead to higher atom flux if the region is too small. Therefore a local maximum of the atom flux occurs. The graph even illustrates that with a sample temperature $T \approx 77K$, a reasonable atom flux can be achieved over a wide range of spot size radius. The discontinuity in atom flux therefore depicts the influence of the heat of fusion.

Fig.(6.10) illustrates, that for large spot-sizes even a target-bath temperature on the order of $T_{target} \approx 77K$ temperature is demanded to investigate the influence of the spot size well, or even to use the spot size variation as trigger for the observation of electron beam induced evaporation. For a sample temperature of 300K a large covered Rb-area itself would be sufficient to evaporate up to 10^{10} trappable atoms/s even at very small beam currents.

The dotted lines give the maximum atom flux for the complete velocity distribution, whereas the solid lines depict the trappable atom flux for 100K (black), 200K (red) and 300K (blue) target-bath temperature.

6.3. Evaporation model and crucial parameter

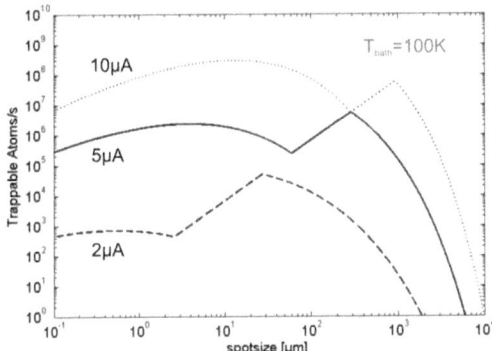

Figure 6.9.: For three dierent beam currents the amount of trappable atoms is shown for a target temperature of $T_{bath} = 100K$.

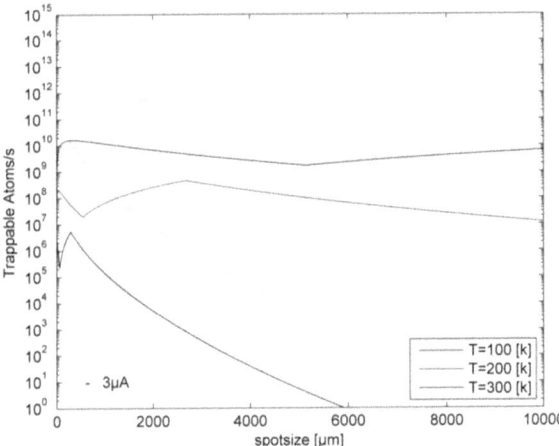

Figure 6.10.: The amount of trappable atoms increases with the target temperature, while for large spot sizes (large targets) even without an electron beam, or a very weak electron beam evaporation occurs due to the low vapor pressure. For lower target temperatures, the influence of the spot size is easier to evaluate

7 Design considerations for an electron beam loaded MOT

7.1. General experimental considerations

The calculations and estimations to achieve a reasonable evaporation (see section 6.3), rise the need for certain demands, while a further implementation into a cryogenic environment must also be considered. In addition, the conceptual setup shown in Fig.(7.1) addresses some experimental questions to be considered such as electrons which cross the MOT-region, the preparation of a cooled Rb-target and the detection of the electron beam, as well as an electron gun/source which is capable of an emission current up to $20\mu A$ while consuming little power to be implemented in a cryogenic environment.

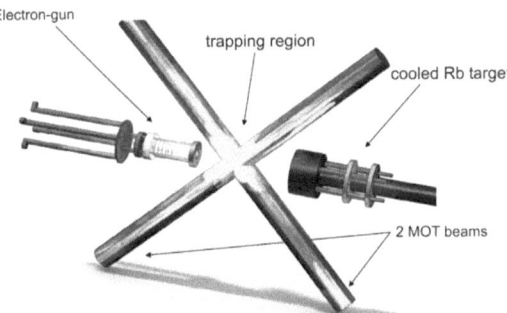

Figure 7.1.: Conceptual setup of an electron beam loaded MOT as it is thought to provide the most optimal and flexible setup

7. Design considerations for an electron beam loaded MOT

7.2. Electron-beam preparation

Electron sources exist in a huge variety of different types. Nevertheless the electron beam source for this experiment has to fulfill certain demands. It should on one hand provide a certain current strength, and a well focused spot size and on the other hand minimize the heat input into the system. In addition the tip should be robust and capable of providing a rather stable current over days/weeks. The demands are summarized in the following

- target current of the electron gun up to $20\mu A$
- compact design of the electron source
- stable and robust emission properties over days/weeks
- total emission power less than 300mW
- therefore emission voltage below 10kV
- capable of a spot size well below 1mm
- focusing system which covers the spot size range up to 1-inch

Known from electron microscopy (EM), there are three different types of field emission tips[1], emitting electrons which are than by a subsequent lens system focused to a spot. The most common tips for transmission electron microscopy (TEM) and secondary-EM (SEM) are therefore thermal field emitters like tungsten tips, or even thermal assisted field emission tips where a LaB crystal emits electrons. Pure field emission tips made of W or PtIr are known from scanning tunneling-EM (STEM) techniques or even from TEM where field emission is used to set free the electrons.
However the emission tip must be followed by a lens-system, capable of focusing the beam to a certain spot size. For the design of such a system, a finite element method calculating the lens-properties and electron trajectories is presented in the following section.

7.2.1. Electron source

As even thermal assisted field emission electrodes are operated with heating powers far beyond the limits satisfying an integration into a cryogenic system (refer to section 4.3), the chosen electron source is a cold field emission tip (see section 5.2.1) either using W or PtIr tips.
In contrast to thermal emitters, field emission tips used in electron microscopes, are just capable of emission currents up to several dozens of nA, but capable of focusing to spot-sizes in the sub-μm range. These small currents are by far not the limit for field emission tips, but currents up to mA can hardly be sustained with cold field emitters.
The principles of field emission are described in section (5.2.1). Regarding to the *Fowler-Nordheim* relation, Eq.(7.2) and the work function of tungsten $= 4.55 eV$, the current from an W-emission tip reaches values up to 100 μA applying rather low voltages up to 1000V assuming a tip radius of $0.2\mu m$. Above a certain electric field strength, the process some

---
[1]Namely: Thermal emitters, thermal assisted field emission and metal field emission tips. In additional there exist carbon nano-tube field emitters for electron guns [58] which occurred in the past years, but are not under investigation for the application in this thesis.

7.2. Electron-beam preparation

when leads to the destruction of the tip, if too high currents are extracted from the surface. Fig.(7.2) illustrates for dierent work functions the emitted current for a tip operated at 6kV. Obviously there is a region in the order of one magnitude in tip radius where even the current diers by 20 orders of magnitude, either leading to the destruction of the emitter or with almost no electrons extracted from the surface.

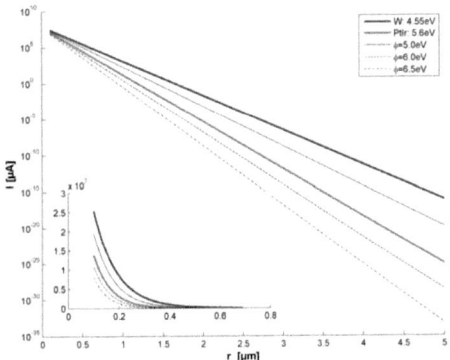

Figure 7.2.: For an emission voltage of 6kV and dierent work-functions, the emission current from a point-like tip is shown, illustrating the strong exponential dependence in the inset.

Regarding to Eq.(5.5), the current density can be written as

$$A_{Emis} \; j(E) = 2 \quad r_{Emis}^2 \; \frac{1.54 \times 10^{-6} E^2}{} \times e^{-\frac{6.83 \times 10^{+7} (\varphi)^{3/2}}{E}} \qquad (7.1)$$

Assuming the tip to be a sphere with radius r, where emission occurs into half-space and the field strength turns into $E = U/r$. Eq.(7.1) can therefore be rewritten

$$I = 2 \; \frac{1.54 \times 10^{-6} U^2}{} \times e^{-\frac{6.83 \times 10^{+7} r \varphi^{3/2}}{U}} \qquad (7.2)$$

From the higher work function of PtIr ($\varphi_{PtIr} \approx 5.6 eV$), it can be seen that less current will be sustained at comparable field strength. Nevertheless PtIr has possibly a twofold advantage. On one hand, PtIr is known from STEM to provide a cheap and simple technique [155], preparing sharp tips just by tearing a wire o with a pilot punch. Additionally, PtIr field emission tips may withstand higher emission currents. From W-tips it is known that they have to be regularly heat-treated to maintain a well formed tip [65]. In other words, PtIr tips may be more robust, easier to fabricate, and easier to handle, whereas W-wires allow for a better preparation of well defined tips.

For the simulation of the lens-system (section 7.2.2) it will be crucial to know the size of the emission region and the tip-geometry quite accurately. Fig.(7.3), therefore shows SEM

7. Design considerations for an electron beam loaded MOT

pictures of a teared-o PtIr tip with wire diameter $d = 250\mu m$, inset a-c) and an incident light microscope picture from an etched W-tip with diameter $d = 250\mu m$. This demonstrates that even with a teared-o PtIr wire, emission tips with radius below $r_{surface} = 1\mu m$ can be realized.

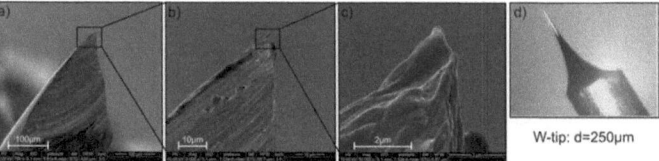

Figure 7.3.: PtIr and W field emission tips as used in the experiment. Left insets a)-c) show a SEM picture of a teared o PtIr wire using a pilot pear. Inset d) shows an image of an etched W-tip with even a point-like tip-region

7.2.2. Focusing lens system

It was pointed out (7.2), that beside the target current, the spot size on the target plays a major role aiming for evaporation of Rb-atoms. Considering a compact electron source, with a small power consumption[2], the experiment demands a field-emission electron-source/lens-system operating at voltages well below 10kV, capable of currents up to $20\mu A$ with a high transfer e ciency from the emission source to the target to keep the overall power consumption low.

The drawback of limited currents using field emission tips can be partly overcome by using a high e cient lens system, loosing as few electrons from the beam as possible, based on a three-anode accelerating lens system rather than with just two focusing electrodes [156]. In principle, neither in microscopy nor in applications using high current electron beams, the e ciency of the lens system in field emission sources plays a major role. This is because in microscopy a tiny, well focused, low current beam is used where the e ciency plays no role, whereas for high power application like EBPVD, more or less field emission tips are completely avoided due to their current limitations.

The type of lens-system

This raises the question how an e cient lens system can be designed. As in principle a lens system for electrons must be capable of focusing a spatially unfocused beam of charged particles, a lens can either use electrostatic or magnetic components, both applying a force on the particles (see section 5.2.2). At this point one should remember that atoms released from the rubidium target will later be trapped in a MOT (refer to Fig.7.1). As in principle the focusing lens system could be built from electromagnetic, and therefore controllable coils, any

---

[2]Both limits the applicable electric potentials in the system as the power consumption goes with $U_{emis} \times I_{emis}$ and the compact size set limits due to breakdown voltages

7.2. Electron-beam preparation

disturbance of the MOT field should be avoided. This would result in the second possibility, namely an electrostatic lens system as it is realized in the setup.

Geometry and electric potentials

The above listed demands for a compact design limit the applied voltages to a certain level, approximately 6kV. This should be the starting point, for further geometrical considerations. As field emission lens systems in microscopes usually are operated at voltages well above 20kV, reaching 100kV[3], the question rises if a lens system for low voltages can somehow have similar properties, when using either an einzel-lens, cylinder- or quadrupole lenses.
This can be answered by a finite element method, calculating the electric fields and potentials in the electron source/lens-system and therefore the trajectories of the electrons.

Emission-tip geometry

Interested in the efficiency of such a lens system, a crucial input parameter for the simulation is the tip geometry. More or less this is the starting condition for the trajectories where small variations in field strength play a major role as shown in Fig.(7.2). For field emission tips it is quite difficult to account for the right tip-geometries in simulations, as for thermal emitters the well-known virtual source diameters of Schottky emitters [157] can be used. Several tip geometries were under investigation, where the geometry and the tip radius were extracted from SEM pictures (Fig.7.3).

7.2.3. Deflection of the e-beam

Adjustment of the beam is shown to be crucial (section 7.3). Additionally the investigations of the spot-size dependent atom flux (section 6.3) demands the beam to be scanned over the target area, to exploit the full target size when the beam is focused. The design basically relys on the calculation due to the movement of electrons in an electric field (section 5.2.2).

7.2.4. Simulation of the electron source/lens-system

Regarding to [158, 159] it turned out, that a good candidate for an realizable configuration could be a three-electrode configuration. Fig.(7.4) therefore illustrates how the electrodes are set up for the simulation. Inset a) shows a screen shot from the simulation program[4] depicting *1st electrode*, *2nd electrode* and *3rd electrode* and the insulation material between the electrodes. In addition to the geometric setup, iso-potential lines are illustrated and some simulated electron trajectories (green) are shown. Inset b) shows a figure taken from [158] which illustrates an archetype to the investigated lens-system.

The beam shape, and the spot size were extensively investigated by varying the geometric shape of the electrodes and finally lead to a design as shown in Fig.(7.4). Attention was paid to optimize for a good efficiency of the lens-system ($\epsilon_{eff} = I_T/I_{emis}$), while the correction of spherical or chromatic aberrations was of less concern.

---
[3] This seems advantageous as the *de Broglie wavelength* decreases with higher electron energy and therefore the achievable resolution increases
[4] COMSOL Multiphysics

7. Design considerations for an electron beam loaded MOT

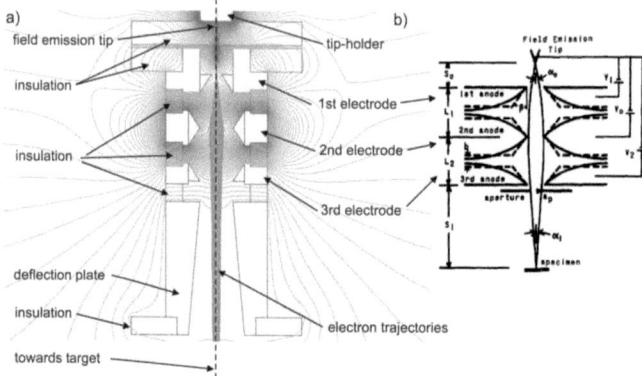

Figure 7.4.: Principle setup of the electrostatic lens system and emission source. Inset a) shows the cylinder-symmetric anode setup with the emission tip, three electrodes and a cylindrical deflection electrode which is sliced in four quarters to maintain arbitrary deflection perpendicular to the vertical axis. Inset b) shows the archetype of the lens system on which the setup is based on. The curved shape of the electrodes can without major drawbacks approximated by conical electrodes as the dashed-line illustrates.

Additional attention was paid to the emission region which was modeled assuming a tip radius similar to that derived from the optical investigation of the used W- and PtIr tips, and the initial transversal velocity component due to the emission process.
For the simulation process itself, the overall boundary conditions were assumed to be the electrode potentials U_1, U_2, U_3 variable for the electrodes $1, 2, 3$ while setting the deflection electrode to $U_D = 0$. The electron kinetic energy belongs to $\Delta E = e \cdot \Delta U_e$, with the target set to zero voltage $\Delta U_e = U_e - 0$.

7.3. Electrons vs. a magneto-optic trap

The chosen setup faces two questions concerning the electron beam and the rubidium MOT. It is worth to think about both, if the e-beam will be perturbed crossing the region of magnetic fields and if there will be an eect of the electron beam crossing the MOT region with the trapped rubidium atoms leading to ionization of trapped atoms. Both questions are examined in the following.

7.3.1. Beam perturbation

Charged particles such as electrons experience a force in a magnetic field (see section 5.2.2). Therefore electrons on their way from the lens-system to the target, will experience the quadrupole magnetic field, crossing the MOT center, as illustrated in Fig.(7.5).

7.3. Electrons vs. a magneto-optic trap

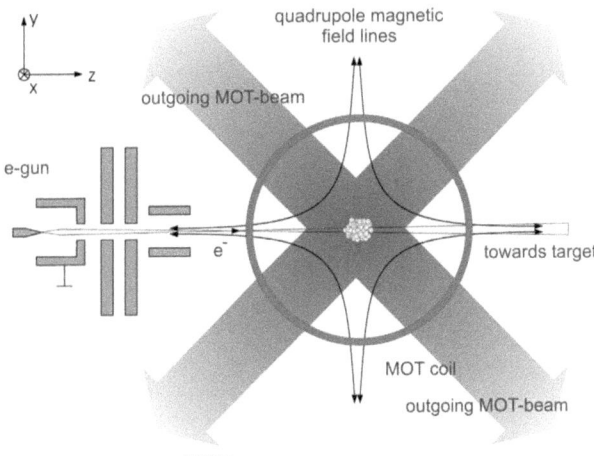

Figure 7.5.: Illustrated crossing of the electron beam with the magnetic field lines of the quadrupole MOT coils. Leaving the sketched electron source, the beam can be adjusted to directly cross the MOT-region on-axis.

Fig.(7.6) shows calculated trajectories of electrons which already illustrate how sensitive the trajectories are to the influence of MOT magnetic field and small initially transversal velocities.
For a typical gradient field of 5G/cm in radial direction and a coil diameter $d_{quad} = 35cm$, the dierent trajectories depict initial osets from the electron beam in x-direction by 1 mm each. Due to the fact that along the zero-perturbation line (red-line) no deflection at all would occur due to the cross product of the velocity v and the quadrupole field B, the initial transversal velocity component is assumed to be 1% of $v_{z0} = \sqrt{2eU_e/m_e}$ with $U_e = 3kV$.
For an initial oset even in the order of 3mm from the longitudinal axis, the electron beam would therefore not reach the sketched target (30mm × 30mm) as shown in Fig.(7.6). This illustrates how important an optimal alignment of the MOT-coils, the electron source and the target is, in order to keep beam adjustment via deflection electrodes as simple as possible.

Fig.(7.7) shows the corresponding transversal velocities of the trajectories illustrated in Fig.(7.6) where abrupt changes in the velocities occur even when the electrons are going to leave the influence of the MOT-coil field (MOT-coils located at $z = 0$). The dierent velocities are given for various transversal oset positions $x_{off} = 0, 1, 2, 3, 4$ and $5mm$, each trajectory starting with $v_{x0} = 1\% \times v_{z0}$.

57

7. Design considerations for an electron beam loaded MOT

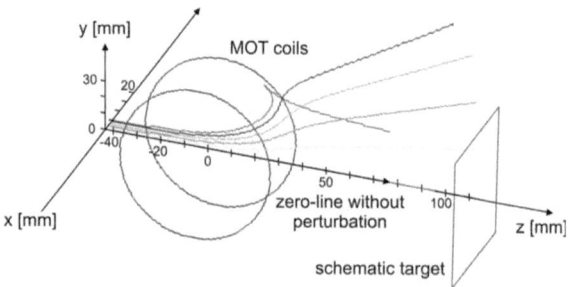

Figure 7.6.: Simulation of different electron trajectories for a typical MOT-field gradient along the propagation direction of $\nabla B = 5G/cm$ from coils in an anti-Helmholtz configuration with diameter $d_{quad} = 35cm$. The trajectories start each with an offset of $x_{off} = 1mm$ from the central axis and an initial tranversal velocity corresponding to $v_x = 1\% \times 3keV$

7.3.2. Electron impact ionization

As the electron beam directly goes along the center line of the setup (see. Fig.7.5), it penetrates the MOT region with the trapped atoms.

Hence this introduces an additional loss-rate in the MOT (see section 5.3.2), as each Rb-atom that collides with an electron, gets ionized and is lost from the trap. This was used both for selectively probing cold trapped atoms [160] and measuring the ionization and excitation cross section between an electron beam and an ultra-cold Rb-target [161] in previous experiments. In addition several experiments were performed which directly measured the electron ionization cross section of rubidium [162, 163] or cesium [164] in a MOT, or even determined the photo-ionization loss rate via this scheme. Therefore the electron impact ionization losses should be considered, whether this is a limiting factor for the loading rate in this setup.

Considering electrons with velocity v_e penetrating an atomic ensemble, the electron impact ionization cross section σ_i is at least dependent on the impinging electron energy [165, 166]. For an electron beam energy of 6kV the ionization cross section can be set to $\sigma_i = 3.5 \times 10^{-17} cm^2$ within confidence [160]. The reaction rate can be written as $\bar{\sigma}_i = n_e \int_v \sigma_i(v_e) v_e f(v) dv = n_e \langle \sigma_i v_e \rangle$. For mono-energetic electrons as concerned above, the number of atoms lost from the trap therefore turns into

$$d\dot{N}_i = N \bar{\sigma}_i = n_e \sigma_i v_e n_{Rb} V_i \qquad (7.3)$$

with the target density n_{Rb}, N the number of atoms trapped in the MOT and the interaction, or ionization volume $V_i = A_{beam} \cdot w$, defined by the electron beam size A_{beam} and

7.3. Electrons vs. a magneto-optic trap

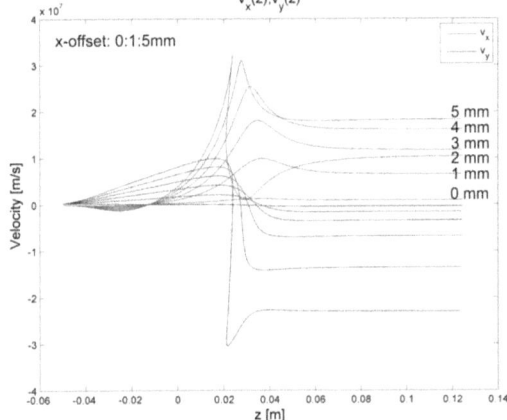

Figure 7.7: Depicted for six different transversal offset positions from $x_{offset} = 0...5mm$, each differing by 1mm, the transversal velocity components are shown for an electron crossing a quadrupole magnetic field with $d_{quad} = 35cm$ and $\partial B/\partial z = 5G/cm$ for $E = 3keV$

the MOT diameter w as the length of the *ionization tube*. The electron density and velocity hence can be described in terms of the beam current I_0 which can be written as

$$I_0 = e \frac{dN_e}{dt} = A_{beam} v_e n_e e \qquad (7.4)$$

with the beam cross section A_{beam}. According to the rate equation Eq.(5.21) for low MOT densities, or Eq.(5.22) in the limit of high atom densities, a new loss term occurs. With Eq.(7.4), this new loss coefficient β_i can be determined from the following equation since $dN_i = \beta_i N$ is valid, and therefore reads as

$$\beta_i N = \frac{6}{w^2} \frac{\sigma_i I_0}{e} N \qquad (7.5)$$

using $n_{Rb} = N/V_{MOT}$ with $V_{MOT} = \frac{w^3}{6}$. The assumption to consider the electron density as constant over the interaction volume V_i therefore holds[5], as the mean free path of the electrons in the MOT-cloud reads as $\lambda_e = \frac{1}{\sigma_i n_{Rb}}$ and is much bigger than the size of the MOT.

Independent of the case of high, or low atom-densities in the MOT this ionization loss mechanism enters as an additional background loss, and the new rate equations would read as

$$\frac{dN}{dt} = L - R' N \qquad (7.6)$$

with $R \rightarrow R' = R + \beta_i = \gamma + \beta_0 n_{Rb} + \beta_i$ for the high density case and

---
[5] Otherwise the drop in beam density must be considered as an exponential decay $n_e(x) = n_e(0)e^{-n_{Rb} x \sigma_i}$

7. Design considerations for an electron beam loaded MOT

$$\frac{dN}{dt} = L - {}'N - {}_0 N^2 \qquad (7.7)$$

with → $' = + {}_i$ for the case of low MOT densities.

For a MOT loaded with 10^9 atoms, assuming a constant density with $n_{Rb} = N_\infty/V_{MOT}$, a spherical diameter of $w = 5mm$ and an electron beam current of $I_0 = 5\mu A$, the relative increase of   is quite small as exhibited by the ratio of the loss rates ${}_i/ \ = 8.35\%$. For the experimental realization this would mean, that the presence of an electron beam, while loading the MOT results in an increase of loading time on the order of less than 10%

7.4. Rubidium Target

The preparation of a proper rubidium target plays a major role in the experimental setup. Rubidium is an alkali metal and therefore chemically reactive especially in combination with water or water vapor. This would therefore restrict the use of larger amounts of pure Rb available in glass cuvettes[6].
Nevertheless the target has to fulfill several demands listed below and following very practical reasons.

- the target should be electrical conductive as it is exposed to the beam current
- the rubidium target should last long enough
- the target should be easy to prepare
- the Rb target should be large enough
- it should be capable of being cooled down even to below 100K

7.4.1. Choice of the target

If therefore a pure alkali target should be avoided the next consideration would be something like a compound. There exist a variety of alkali metal salts, e.g RbCl, which can be oxidized using Ca, Na, Mg, or Zn, following a reaction equation like

$$RbCl + Na \xrightarrow{T_s} Rb + NaCl \qquad (7.8)$$

or compounds which can easily be handled[7]. Nevertheless this process may occur at temperatures, higher than necessary for the desorption of RbCl molecules which would not lead to the desired desorption of atomic rubidium.
Known from alkali metal dispenser [123], there is another possibility, much more promising. With a mixture of Zr and Al, $Rb_2(CrO_4)$ is turned into pure atomic rubidium vapor, where in addition several oxides are released, following the reaction equation:

$$Rb_2CrO_4 + x\,Zr + y\,Al \xrightarrow{823K} 2Rb + Cr_2O_3 + ZrO_2 + Al_2O_3 \qquad (7.9)$$

---
[6]NEYCO, Paris - France.
[7]Rb_2CO_3, and Rb_2O in form of pellets, tablets and powder, *American Elements*, Los Angeles, CA USA.

7.4. Rubidium Target

Thus it would be a choice to prepare a target surface [131] by evaporating Rb from an resistive heated dispenser[8], located very close to the target.
As the partial pressure of evaporated Rb, released from the strongly driven dispenser next to the target surface, would be much higher than the vapor pressure of Rb[9] at the target temperature, this would result in nothing else than a coating of the target surface with a film of atomic rubidium, as well as with all the other species evaporated from the dispenser, following the rate-equation for particles right above the surface as

$$\frac{dN}{dt} = A_{surf} \times \frac{p_p - p'_v}{\sqrt{2mk_BT}} \qquad (7.10)$$

with the dierence between partial pressure p_p and the vapor-pressure p'_v of the condensed evaporant, which is valid both for evaporation and condensation of vapor next to a surface. The rate of condensation of a Rb-vapor next to the surface is even enhanced, if the target is cooled below room-temperature.
The mean free path of rubidium $_{Rb} \propto \frac{1}{n_{Rb}d^2}$, indirect proportional to particle density and particle size d^2 at pressures below $10^{-9}mbar$ is therefore large enough that the atoms can travel from the dispenser to the surface.

7.4.2. Target consistency

If the target is considered to be prepared the above mentioned way, the coated Rb layer[10] must have minimal thickness to satisfy at least the *one-shot ratio* defined by the beam waist of the impinging electron beam, and the thickness of the Rb-layer.
Regarding to the mass flow calculation shown in section (6.3.3), the estimation results in the following demands for the thickness d_{min} and capabilities, if about 10^8 trappable atoms should be evaporated from a target surface with $A_{target} = 30mm^2$.

- $w_{beam} = 100\mu m \Rightarrow d_{min} = 30nm$

- $w_{beam} = 10\mu m \Rightarrow d_{min} = 3\mu m$

- for $d = 3\mu m$ $A = 30mm^2 \Rightarrow$ target lasts for 80.000 cycles

- for $t_{evap} = 3s$ switch on in $t_{cycle} = 30s$, target lasts for 1 month for continuous operation

---

[8]several investigations of ESD phenomena for alkali metal atoms also use this technique to prepare the target
[9]See reference [140] for Rb-vapor pressure curve.
[10]For further considerations the target is now assumed to consist of a layer of atomic rubidium even if other elements are incorporated in the layer

8 Setup of an electron-beam loaded MOT

8.1. Laser system and MOT optics

The laser system is described in detail in [154]. It consists of the basic parts necessary for achieving a simple setup for trapping cold ^{87}Rb atoms in a MOT. In addition a three laser system is chosen where the high power *cooling-laser* is stabilized by a frequency-oset lock to a reference laser. Thus the cooler laser is capable to cover a detuning range of up to ± $70MHz$ around the cooling transition without the need for shifting the frequency by the use of an acusto-optic modulator (AOM).

- A frequency stabilized diode laser at the ^{87}Rb cooling transition $5^2S_{1/2}|F=2\rangle \rightarrow 5^2P_{3/2}|F'=3\rangle$ which acts as an reference laser

- A master laser with an additional Tapered Amplifier locked via a frequency-oset (FO) lock to the reference laser

- A distributed feedback diode laser frequency stabilized with an frequency modulation (FM) lock at the ^{87}Rb repumping transition $5^2S_{1/2}|F=1\rangle \rightarrow 5^2P_{3/2}|F'=2\rangle$

- two AOMs, one each in the cooler- and the repumper-path acting as a fast switch for shutting o the laser light

- a telescope to enlarge the overlayed cooler- and repumper-light to a diameter of 1 inch.

Fig.(8.1) shows the setup of the MOT optics. Both, the cooler- and the repumper-light ($\approx 4mW$ at the MOT-chamber) are out coupled from an optical fiber and focused onto an AOM each. The beams are overlayed in a polarizing beam splitter and pass a telescope which enlarges the laser beam up to 1 inch diameter. Furthermore the light is split into three beams, each re-troreflected after passing the vacuum-chamber (see section 8.4) with a total light power of 71.4 ± $0.1mW$. Six quater-wave-plates in addition allow to adjust the polarization of the ingoing and outgoing beams (see section 5.3.2).

8. Setup of an electron-beam loaded MOT

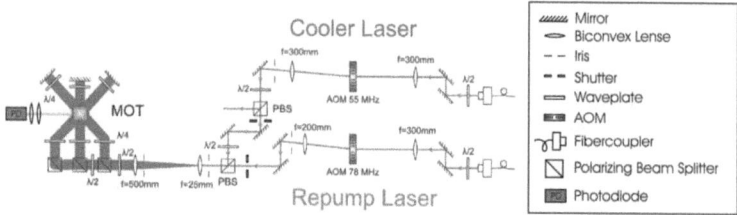

Figure 8.1.: Laser setup for loading a MOT as used in the e-beam experiment. The figure illustrates the two out coupled beams focused onto AOMs and overlaid in PBS-cube. A subsequent telescope enlarges the beams and three cube splits the laser-beam into three equally beams that are retro-reflected to obtain a MOT.

8.2. Electron-gun

Two dierent types of electron-guns (e-guns) were constructed and tested in the experimental setup, as it turned out that the prototype did not lead to the desired results. The e-gun presented in the following is from the 2nd approach and was operated with dierent emission tips.

8.2.1. Preparation of the field emission tips

As previously described (section 7.2), the tips must fulfill certain demands. Therefore an etched tungsten[1] tip and a teared-o PtIr [2] tip were applied as emitters for the e-gun, both with a wire diameter of $250\mu m$.

The generation of PtIr tips can easily be done by just tearing o a wire with a cutter plier under a flat angle. Nevertheless every tip must be proved either with the naked eye before mounting, and with the SEM as shown in Fig.(7.3) after dismounting it from the experiment. For etching W-tips extensive literature [64, 167] exists and with a number of trials the desired point-like tips could be produced, following the receipt described in [154].

8.2.2. Emission properties

The diverse emission properties mainly result due to the dierent work function of W ($_W \approx$ $4.55eV$)[3] and PtIr with its much higher value of $_{PtIr} = 5.65eV$. Hence PtIr demands higher emission voltages on the order of $-3kV... - 6kV$ to reach the same currents as with tungsten at $-0.4... - 0.8kV$. The emitted tip-current is critical to the applied electric field, and for point-like tips it could happen, that high emission currents change the tip-geometry.

---
[1] W 0.25mm diameter, 99.9%, Goodfellow
[2] Pt/Ir 0.25 diameter, 90/10%, Goodfellow
[3] This is a typical value for the prepared emission tips.

8.2. Electron-gun

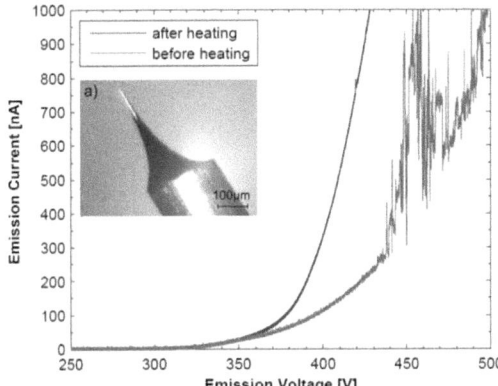

Figure 8.2: Emission characteristic of an etched W-tip before and after heating. The fluctuating current origins from the surface being damaged by high emission currents which cause much worse emission properties after an emission arc occurred. After heating the tip with 1W power at least for 20s, the tip performance increases as the damaged tip reforms. The graph is taken from [154]. Inset a) depicts a picture of an etched W-tip taken with an optical microscope

If the tip is operated for a while at high currents, the emission voltage for constant current tends to increase. This can be overcome by flash-heating the tip[4]. Therefore the tip-surface after an extensive emission process gets smoother, possibly with the drawback of a bigger curvature. Fig.8.2 inset a) shows a W-tip where emission properties can extensively be increased heating the tip for 20s at a power of $\approx 1W$, whereas inset b) shows a smoother but smaller emission current after flash-heating.

Furthermore, both tips are capable of sustaining an pulsed emission current of $I_e > 15\mu A$ ($t_{pulse}/t_{pause} = 2s/1s$), lasting for at least more than 2000 cycles. PtIr tips with the higher work-function also allows to go for currents up to $110\mu A$.

8.2.3. Focusing electrodes

The lens-system consists of three focusing lenses as illustrated in Fig.(8.3), namely electrode 1-3. The outer diameter of the ring shaped electrodes is 18mm for all three of them, while the inside diameter is bigger for electrode 2 and 3 in order to do not restrict the maximum numerical aperture. The electrodes are connected to the high voltage power supplies through a Cu-wire plugged in a bore-hole at the outer circumference (see Fig. 8.4) of the electrode[5]. The electrodes are made of OFHC-Cu[6] and surface-treated as described below.

8.2.4. Deflection plates

The four conical deflection sections, sliced from one cylindrical electrode allow the beam to be bended in x- and y-direction, respectively in the complete x-y plane. Applying a voltage dierence $_D$ between $\pm 175V$ allows maximum deflection up to $\pm 15mm$ at the target. This

---
[4]spot-welded 2pin-tips which were heated for several seconds with currents up to 7A
[5]The bore-hole is slightly less in diameter than the bare Cu-wire which allows the wire to be plugged in.
[6]Oxygen-Free High thermal conductivity Copper

8. Setup of an electron-beam loaded MOT

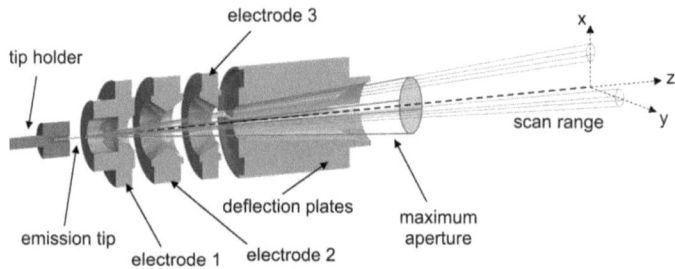

Figure 8.3.: A cut through the electron gun is shown, illustrating the three high-voltage electrodes, the tip-holder and the four conical deflection plates. It illustrates the maximum aperture of the focusing lens-system and the deflection capability of the system.

corresponds to $_{x,y} = 86\mu m/V$ at the target or to $\approx 0.23 mrad/V$. The deflection plates are just sliced without additional gluing and kept in position by the ring-shaped clearance at the electrode bottom.

8.2.5. Mounted E-gun

An overview of the assembled e-gun is shown in Fig.(8.4). The electrodes are kept together by a top- and a bottom macor[7] cap sustained by four threaded steel rods with nuts ontop that can be tightened. The inter-electrode spacing is maintained by 3 macor-discs which fits into the electrode clearings (see Fig.8.3). The picture also shows the connection wires (insulated Cu) which are plugged into bore-holes at the outer circumference of the electrodes, some of them connected at the bottom macor cap, others carried down to the flange which provides 9 HV-feed-throughs. Four of those feed-throughs maintain the four Cu-rods which carry the bottom macor cap. The deflection plates are connected via the four top-connection screws, the other wires are carried down to the flange through four pin-holes.
The tip holder (not seen in the overview) is directly connected with the central feed-through maintaining the emission voltage.
All electrodes are electro-polished and coated with a $1\mu m$ thick Au-layer on top of a $1\mu m$ thick silver layer. The overall length of the electron gun from the bottom macor cap to the top macor cap is $L \approx 85mm$ with an overall diameter of $d_{top} = 29mm$ and $d_{bottom} = 50mm$. Find the detailed construction drawings in Appendix (G).

The compact design of the electron-gun limits the maximum applicable potentials as otherwise arcs and electric breakthrough would occur. For vacuum a breakthrough voltage of $10kV/mm$ is quite a common value, whereas the creep-distance between two electrodes over the macor surface has to considered to limit the breakthrough voltage to about $1kV/mm$.

---
[7]Machineable glass-ceramic, Corning Incorporated

8.2. Electron-gun

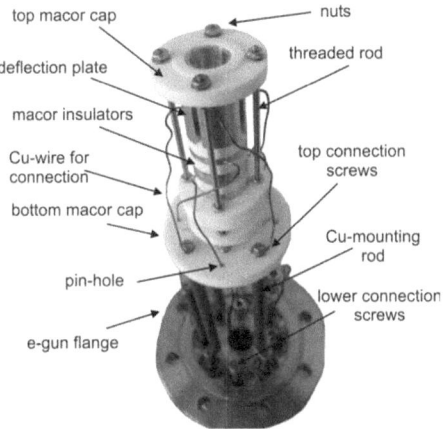

Figure 8.4.: Overview of the assembled electron gun mounted on a CF63 flange

Tab.(8.1) shows typical applied voltages at the emission tip and the electrodes, and gives the maximum applicable potential limited by the breakthrough voltage.

electrode	typical applied voltage	maximum potential dierence
	$U_i = \ - GND$ [kV]	Δ_{max} [kV]
E_1	-1.0...+2.0	3.0 to E_2
E_1	-	2.0 to threaded rod
E_2	-6.0...+3.0	3.0 to E_3
E_3	-3.0...+2.0	3.5 to U_D
tip	-6.0...0	7.5 to threaded rod
tip	-	8.0 to E_1
U_D	-0.2...+0.2	1.5 to threaded rod

Table 8.1.: Maximum and typical operating potentials applied at the e-gun electrodes

All parts of the electron-gun are carefully cleaned and prepared for UHV but not especially baked out, as the chamber itself is heated up to 200° every time after rebuilding[8].

---
[8]During preparation of the vacuum chamber the e-gun itself is already mounted inside the chamber, and baked out by thermal conduction through the e-gun mounting.

8. Setup of an electron-beam loaded MOT

8.3. E-gun targets

Following the above given considerations (see section 7.4) different target types were investigated, reaching from very trivial setups, to water- and LN_2-cooled targets. For characterization of the electron gun, a modified phosphor screen was attached to the chamber instead of evaporation targets as described in the following.

8.3.1. Nitrogen cooled targets

Most of the results shown in chapter (9), were obtained using a liquid nitrogen (LN_2) cooled Cu-target prepared with two Rb containing alkali metal dispenser (AMD) in front, Fig.(8.5) inset a). The LN_2-cold finger (top-loaded) and the electrical connectors are assembled in a CF63 4-way cross, finally attached to the vacuum-chamber. Detailed drawings can be found in Appendix (G).

To prepare the cooled target, both dispenser are driven with 5A for 60s, while most of the Rb sticks on the Cu-surface as the partial pressure of Rb is 10^{-15}mbar at 77K [140]. To get a measure of the target temperature, a Pt100 sensor[9] is connected to the Cu-pot. To measure the target current, a bare Cu-wire is plugged into a bore-hole at the backside of the Cu-pot and connected via a feedthrough to the experimental control.

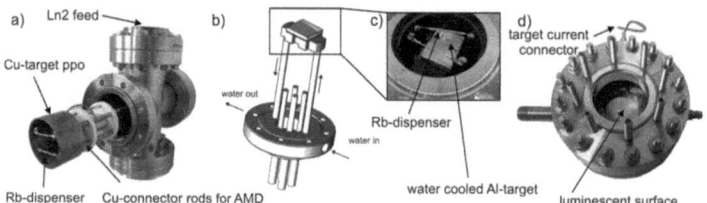

Figure 8.5.: Overview of the different e-gun target, showing the liquid nitrogen cooled Cu-target with 2 AMD in front, inset a). A water-cooled Al-target setup is shown in inset b) and c) with a thick coating on top after dismantling the target. Inset d) shows the modified phosphor screen setup with a stack of adapter flanges.

8.3.2. Standard targets

For some measurmenets a simpler water-cooled Al-target is used, prepared in the same way as the nitrogen target, coating it with two near-by AMDs. With an attached water chiller, the target temperature can be cooled down to $-2°$. Fig.(8.5), inset b) shows the principal setup of the target flange, while inset c) shows a thick coating on the Al-surface after dismantling the target from the chamber. Again the target current can be measured via Cu-wire connected at the backside of the Al-plate. To avoid a short to ground, the Al-plate is insulated from the

---

[9]Gauged down to $T = 4K$

steel-mounting via a thin Kapton foil glued to the mounting with thermal conductive epoxy[10].

8.3.3. Phosphor-screen

To characterize the beam profile, the deflection- and focusing properties, a simple method is to use a luminescent phosphor-screen, commonly found in electron optics. The implemented phosphor-screen was attached to a micro-channel plate detector which was removed. Existing electrical connections were removed, while a signal line to the coated window (luminescend surface) was established. This allows to measure the impinging electron current, while a camera was used to take pictures of the electron spot on the screen[11]. Inset d) of Fig.(8.5) shows the phosphor screen (the luminescent coating on a thick window) with a stack of CF100 adapter flanges which can be connected to the vacuum-chamber.

8.4. Vacuum-system for the e-beam MOT

The room-temperature vacuum chamber for the e-beam MOT setup consists of the five-axis main-chamber with three orthogonal axis maintaining a CF63 flange and a CF40 cross, tilted by 45° maintaining two of the MOT a axis. The chamber can be baked out after every rebuilding and is pumped through a turbo-molecular pump and a 200l/s ion-getter pump with an implemented Ti-sublimation pump. Two CF63 adapter tubes on both sides of the horizontal axis sustain the electron-gun flange on one side and the target flange on the other side.

8.5. Experimental setup

The MOT-coils are directly wounded onto the main-chamber while compensation coils are omitted. They produce a quadrupole field at the MOT-center with 2.7G/cm/A coil-current described elsewhere in detail [154].

The detailed setup is illustrated and depicted in Fig.(8.6) inset a), which sketches the inner life of the vacuum-system shown in inset b). It shows the LN_2 cooled target flange assembled in a CF63 4-way cross attached to the chamber. This 4-way cross can easily be replaced and alternatively equipped with the water-cooled standard target or the phosphor-screen (see section 8.3). The number of trapped atoms in the MOT is further characterized by resonance fluorescence detection [81]. The used photodiode[12] is attached to an collimation tube with a total focal length of 235mm.

The experiment is controlled via a real-time processing system[13]. In the implemented way, it allows to update analog and digital output-signals even within $20\mu s$. The target current, and the emission current as well as the chamber pressure[14] are continuously monitored via

---
[10] Find more details on the epoxy in Appendix (A.7) and (D)
[11] The measured electron beam spot size was directly determined by the fluorescence on the screen. Magnifying imaging behaviors of the screen were not taken into account
[12] DET110 *Thorlabs*: $\eta_{eff} = 0.53A/W$
[13] ADWin-PROII, *Jäger Messtechnik, D-64653 Lorsch*
[14] UHV-24p *Varian* hot-cathode ion gauge

8. Setup of an electron-beam loaded MOT

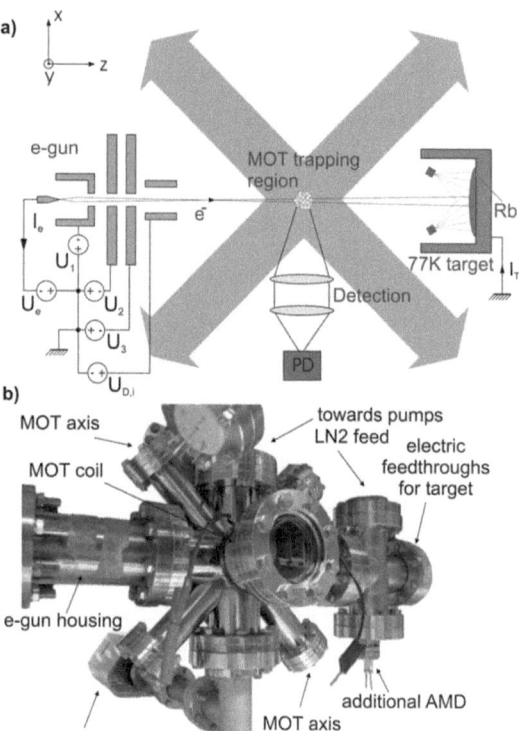

Figure 8.6: Inset a) illustrates the schematic of the cryogenic e-beam driven alkali atom source for loading a MOT: electrons emerges from the cold field emission source, and the beam leaves the electron-gun, crosses the trapping region with the crossed laser beams and hits a liquid nitrogen cooled rubidium target. Two alkali metal dispensers are used to prepare an alkali metal layer on the Cu-surface. Released ^{87}Rb atoms are than loaded into the MOT and the fluorescence of the trapped atoms is measured with a photodiode (PD).
Inset b) shows the experimental setup with the main-chamber and the crossed MOT-axis. The e-gun and the target are located at the horizontal axis while the mass spectrometer and the pumps are mounted on the vertical CF63 flanges.

8.5. Experimental setup

an independent *Labview*-based measurement system, while the partial pressure of different species is recorded with a quadrupole mass spectrometer[15]

---

[15]Mass spectrometer: Quadstar 422. *Balzers*

9 Results: The electron-beam-MOT

The following chapter presents both, emission properties of the novel high-efficient electron gun, as well as results for the electron beam driven atom source. Part I then finishes with a model that describes electron stimulated desorption, as the measurement results indicate, that this is the underlying effect for desorbing atoms loaded in the MOT, with the use of an electron beam.

9.1. High efficient electron-gun

The design goals for the electron-gun (see section 7.2) finally lead to the construction of the highly efficient [1] electron-gun (section 8.2). Capable of a transfer efficiency η_{trans} up to 30%, the cold field emission driven electron gun is capable of providing up to $110\mu A$ beam current focused down to the sub-mm region at beam energies of just $6 keV$.

Focusing behavior

The transfer efficiency $\eta_{trans} = I_T/I_e$ can be measured, as it is defined by monitoring both, the emission- and the target-current. Hence the dependence on the focusing voltage U_2 provides a deeper insight on the focusing behavior, nevertheless also U_1 and U_3 influence the focus and efficiency.

Fig.(9.1) shows the beam waist measured on the phosphor screen (see section 8.3), with the measured intensity distribution shown in the inset. The beam can be focused to the sub-mm regime with a minimum waist at an optimal focusing voltage U_2.

Beam transfer-efficiency

Nevertheless, one important question for the design considerations was, whether the electron beam can be steered without perturbation through the magnetic field of the quadrupole MOT-coil. Fig.(9.2) shows a nearly similar behavior as found for $I_{MOT} = 0$, but with a slightly decreased efficiency $\eta_{trans} \approx 21\%$ for optimal focusing voltage, even at a magnetic

---
[1] Transfer efficiency between emission current I_e and target current I_T

9. Results: The electron-beam-MOT

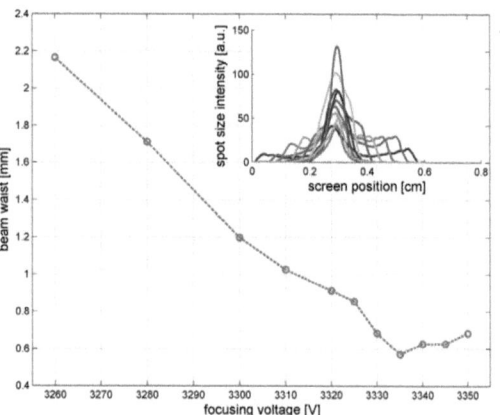

Figure 9.1:
Focusing behavior of the electron-gun in dependence of the focusing voltage applied at the 2nd electrode, without the MOT-field. The inset in the right shows the intensity distribution, measured at the phosphor-screen attached opposite of the electron-gun, from which the FWHM, the Gaussian radius of the spot size was determined.

quadrupole gradient of $B_{MOT} = 13.5 G/cm$. In addition it turned out, that the maximum efficiency does not coincidence with the smallest spot size. Fig.(9.2) illustrates that a careful beam adjustment with the deflection plates allows to stear the electron beam through the quadrupole field without crucial perturbence.

9.2. Cold atom source

Evidence for atoms to be released from the liquid nitrogen cooled target is shown in Fig.(9.3). After preparation of the target (see section 8.3.1), the e-gun is switched on, releasing a beam towards the target surface, while simultaneously monitoring the trapped atoms as described below. The graph depicts clearly that the increase of trapped atoms occurs in the experimental cycle[2] when the beam is for the first time switched on.

The measurement cycle relies and four phases I-IV, while in every phase the atom-number in the MOT is evaluated using the method of resonance fluorescence detection [81]. At the end of each experimental phase the MOT laser frequency is ramped from $= -18 MHz$ within 2ms over resonance to a blue-detuning of $= +5 MHz$. The inset of Fig.(9.4) shows an exemplary fluorescence signal captured with the photodiode. Hence the relative height of the peak is a measure for the atoms trapped in the MOT.

In phase I, the MOT is loaded while the electron beam is switched on, in order to get a reference measure of the background as shown in Fig.(9.5, 9.6, 9.7).

In phase II the electron beam is switched on and the desorbed atoms can be trapped in

---
[2]Indicated by the global counts which gives the number of experimental cycles.

9.2. Cold atom source

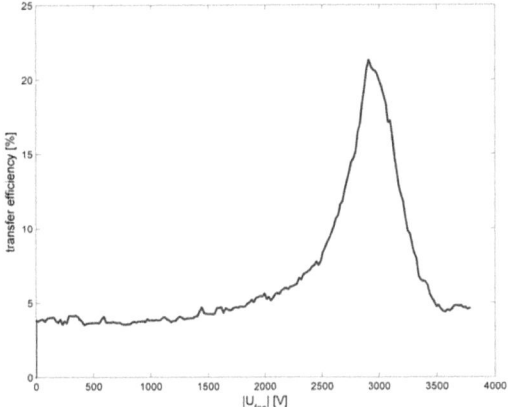

Figure 9.2.: Measured transfer efficiency, which gives the ratio of emitted current I_e and measured target current I_T for different negative focusing voltages applied at electrode E_2, while the MOT-quadrupole field is on with $B_{MOT} = 13.5 G/cm$. The maximum peak current at $U_2 = -2.90 kV$ is $\approx 2.1 \mu A$. It should be mentioned that maximum transfer efficiency does not correspond with smallest target spot size.

the MOT. To switch off the electron beam after phase II the emission voltage is canceled by disabling the HV switch shown in Fig.(9.4).

During phase III, atoms trapped from the remaining background gas are detected while the e-gun is off. This allows to get a feeling how fast the background pressure gets back to the residual pressure after switching of the e-gun.

After a pause, caused by the need of ramping up the suppression voltage, phase IV measures the trapped atoms if the e-gun is switched on, but blocking the beam. This is achieved by applying a large negative voltage at the electrode 3, which is ramped up right before phase IV starts. In phase IV the beam is therefore not hitting the target, even though a small leakage current may remain. This phase allows to assess if the trapped ^{87}Rb atoms origins from the target rather being released from other surfaces accidentally hit by electrons.

Desorbed atoms

Varying the target current via the emission voltage U_e regarding to *Nordheim-Fowler*, Eq.(7.2), a linear increase in atom number is observed in phase II (N_{II}), see Fig.(9.5). The electron-gun is operated with a PtIr tip biasing the first electrode with $U_1 = -500V$ while U_e is varied from $-2.5kV$ to $-3.0kV$. The focusing voltage is set to $-440V$. As each phase lasts 1500ms the loading rate of atoms into the MOT can be determined the following way.

9. Results: The electron-beam-MOT

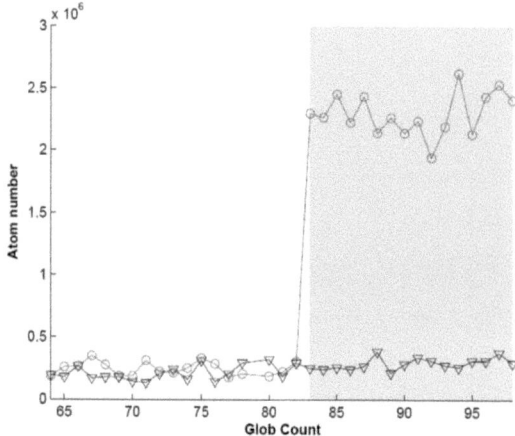

Figure 9.3.: Atom-number trapped in the MOT and counted via fluorescence detection for various cycles before, and after switching on the electron-gun, directed onto the prepared Rb-target. The x-axis depict the glob-counts where each tick indicates a separate experimental cycle.

The loading time constant for loading the MOT, turned out to be on the order of 10s, measured in a previous run[3]. Hence for an experimental loading time of $t_L = 1500ms$, the loading rate of the MOT can be deduced by $R_{II} = N_{II}/1.5s$ and the axis of Fig.(9.5) can easily be scaled to $N_{II} \to R_{II}$. The loading yield Y_D which describes the desorption of atoms and the trapping in the MOT with the target current I_T therefore is directly proportional to the slope of the dashed line in Fig.(9.5) where $Y_D = 5.1 \times 10^5 atoms/s/\mu A$.

In a typical experiment 3×10^6 atoms are trapped in the MOT at target currents of $2.8 \mu A$ and for longer loading times $t_L = 2.5s$ up to 4.5×10^6 trapped atoms can be achieved.

Unfocused desorption

Operated with a totally unfocused electron beam with a beam size on the order of cm^2 and target currents well below $1 \mu A$ as shown in Fig.(9.6), by setting the focusing voltage to zero, even atoms can be trapped in the MOT. It is interesting that obviously two focusing regimes can obtain local maxima in trapped atom number. First, while the beam is unfocused which leads to a small target current, as the e ciency $_{trans}$ is low and the impinged area large, and second for a rather focused beam with a higher target current I_T but a smaller beam spot size.

---

[3]Fitting the governing exponential law to the loading curve, according to Eq.(5.22), the characteristic time constant τ_L can be deduced following $N(t) = N_\infty (1 - e^{-\frac{t}{L}})$

9.2. Cold atom source

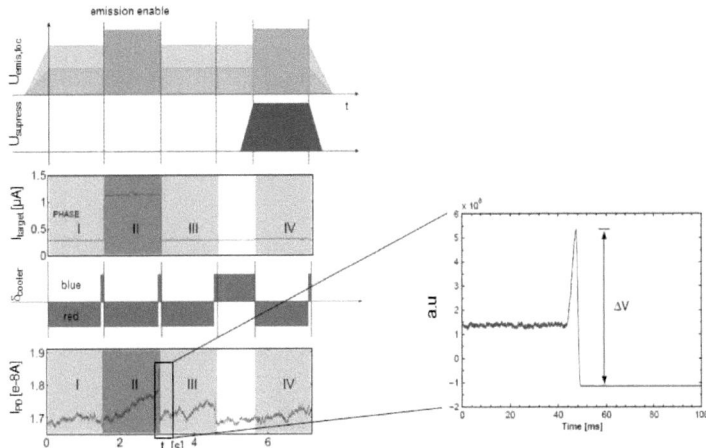

Figure 9.4.: Specific experimental cycle with phase I-IV, which exhibits the typical corresponding target current and atom number trapped in the MOT measured via fluorescence detection.

Even with a W-emitter operated at $U_e = 650V$ and an unfocused beam with $I_T \approx 330nA$, a loading rate of $7.3 \times 10^5 atoms/s$ could be achieved during 1.5s loading time. A reference measure loading a MOT at the same background pressure of 1.2×10^{-10} mbar by a conventional AMD yields a loading rate of $\approx 1.9 \times 10^7$ atoms/s, using a resistive power of 18W. When scaled to the applied beam power of $\approx 200\mu W$, the electron beam driven atom-source leads to a loading rate per input power 1000 times bigger than for a conventional loaded MOT.

Compared to the calculation done in section (6.3) this rather indicates a dierent process than electron beam induced evaporation as the spot size a) is much bigger, and b) the deposited power density is much smaller than theoretically demanded. While even the linear behavior in the number of desorbed and trapped atoms N_{II} with target current I_T tends to be conflictive with the behavior described by Eq.(6.27), another fact indicates a dierent eect.

Target temperatures

Indicated by Fig.(6.10) the number of trapped atoms should vary by at least three orders of magnitude for spot sizes in the order of $r_{spot} \approx 1mm$ or below, for higher target temperatures. A measurement at a cold and hot target as shown in Fig.(6.10) shows a dierent result. The number of trapped atoms for both runs shows a linear behavior, while diering by less than a factor of two in the number of trapped atoms at $I_T = 1.4\mu A$. It tends, that for very small target currents on the order of $100nA$ or less, the target temperature plays a less

77

9. Results: The electron-beam-MOT

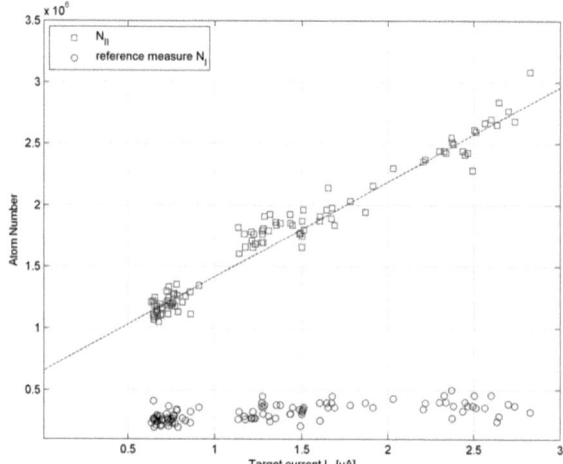

Figure 9.5.: Atoms desorbed (□) from the Rb-target measured in phase II, increasing with electron-beam current in comparison with atoms measured in phase I (o) while the electorn-gun is switched o .

important role as for higher currents. Hence, this tends to a process e ected much less by target temperature than electron stimulated evaporation tends to be.

9.3. Model for Electron Stimulated Desorption

The above given results, indicates that thermal desorption due to electron beam heating can not be the relying process setting free Rb-atoms. Nevertheless as shown in section (6), and according to [128, 129, 130, 131], ESD is obviously the mechanism that can be considered to be responsible for the above given observations. Compared to EBPVD, ESD does not demand high electron beam densities as the e ect is almost based on auger induced desorption of alkali metals, desorbed on an oxygen layer. An incident electron therefore creates in the 2s oxygen level a core-hole which triggers an intra-atomic Auger-decay. This leads to a subsequent neutralization of one positive alkali metal ion. Hence if the positive oxygen ion can capture electrons from the substrate to achieve a negative charge state again, the alkali metal atom is repelled and desorbed from the surface as a neutral atom.

The e ect of ESD is applicable even to molecules [168] from solid inert gas surfaces, and molecules desorbed from metal surfaces [4]. Hence the molecular desorption yield derived in

---
[4]The desorption of molecules from metal surfaces, even at cryogenic temperatures turned out to be a big

9.3. Model for Electron Stimulated Desorption

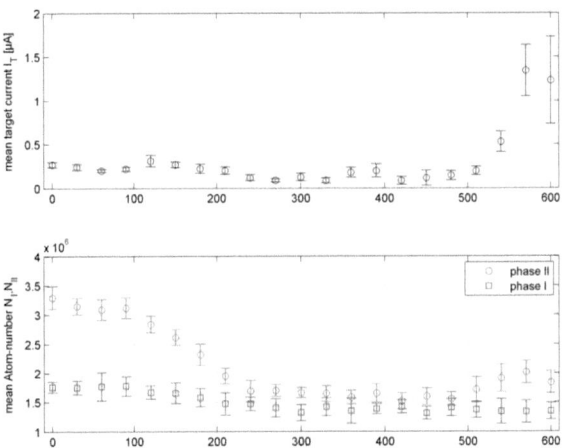

Figure 9.6.: Target current and mean atom number from phase I and II varying with the focusing voltage and hence the spot-size of the impinging electron beam

[170] can in principle also be applied for the presented results as it relies on the investigation of increasing partial pressures only.

If a mass spectrometer records the partial pressure above the target surface described by

$$p_i = n_i k_B T \qquad (9.1)$$

the pressure p_i of the species i in a vacuum chamber with volume V than reads as

$$V \frac{\partial p_i}{\partial t} = \dot{Q}_i - p_i S_i \qquad (9.2)$$

where a steady-state pressure after an infinite time is reached through the vacuum pumps that pumps species i with S_i $[m^3/s]$ and the desorbed gas load $\dot{Q}_i$ $[Pa m^3/s]$, leading to

$$\dot{Q}_{i0} = p_{i0} S_{i0} \qquad (9.3)$$

If the gas load due to a physical desorption process is suddenly increased, a new steady-state pressure will be observed which reads after an infinite time as

---

issue at the particle accelerators at CERN, as this lead to an pressure increase [169] caused by ESD and PSD

9. Results: The electron-beam-MOT

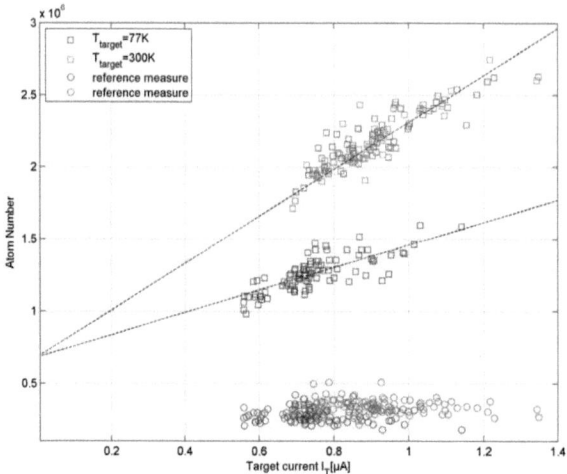

Figure 9.7.: Comparison of the atom-number and the desorption yield for dierent target temperatures

$$\dot{Q}_{i1} = p_{i1} S_{i0} \qquad (9.4)$$

This increase in gas load can now be related to a desorption yield caused by a certain process and reads than as

$$\Delta Q = (p_{i1} - p_{i0}) \times S_{i0} \qquad (9.5)$$

With Eq.(9.1) and Eq.(9.2), ΔQ turns into

$$\Delta Q \propto \dot{n}_i \times k_B T \qquad (9.6)$$

and an observable desorption yield Y_D can be defined as

$$Y_D \propto {}_i\frac{I_T}{e} \times k_B T_T \qquad (9.7)$$

Hence, all the intrinsic desorption parameter enter in ${}_i$. The desorption/loading yield Y_D therefore depends on the target temperature T_T and the number of atoms per target current dN/dI_T, and exhibits a linear behavior as depicted in Fig.(9.7) and Fig.(9.5). The number of trapped atoms N_{II}, scaled to the target current therefore shows even a steeper slope for higher target temperatures as given by

80

9.3. Model for Electron Stimulated Desorption

$$\frac{N_{des}}{I_T} = \int_0^t Y_D(T_T)dt \qquad (9.8)$$

Nevertheless the desorption constant σ_i also may depend slightly on T_T [171, 172], while studies on the influence of surface coverage and electron energy also tend to affect the desorption yield Y_D. Comparison with literature [173] shows that even for an electron target density much smaller than the adatom density Eq.(9.7) holds[5]. Even electron energies above 400eV and up to the range of 6keV do not influence the desorption coefficient σ_i for alkali metals much.

Indication for ESD by partial pressure measurement

Even the experimental setup includes a quadrupole mass analyzer (QMA)[6], it is not trivial to calibrate it for rubidium. Hence the evaluation and further investigation of Y_D and σ_{Rb} could not be done. Nevertheless Eq.(9.6) is the strongest evidence for ESD, while the measurement of the relative increase in QMA-current as shown in Fig.(9.8) is another indication that desorbed Rb is due to ESD rather than thermal desorption, as the partial pressures of the molecules[7] H_2, N_2, CO_2, ^{85}Rb and ^{85}Rb strongly increase linear over several orders of magnitude if the electron gun is switched on.

---

[5] With a target current of $1\mu A/cm^2 \approx 10^{13} electrons/cm^2$ and a typical adatom density of $10^{15}/cm^2$ this condition is easily fulfilled

[6] The Quadstar mass-spectrometer ranges up to $m/q = 100$ and was not calibrated as this is not trivial for rubidium

[7] The QMA detects the ratio m/q and for m/q=2,14,44,85,87 the common associated molecules in this setup are therefore listed above

9. Results: The electron-beam-MOT

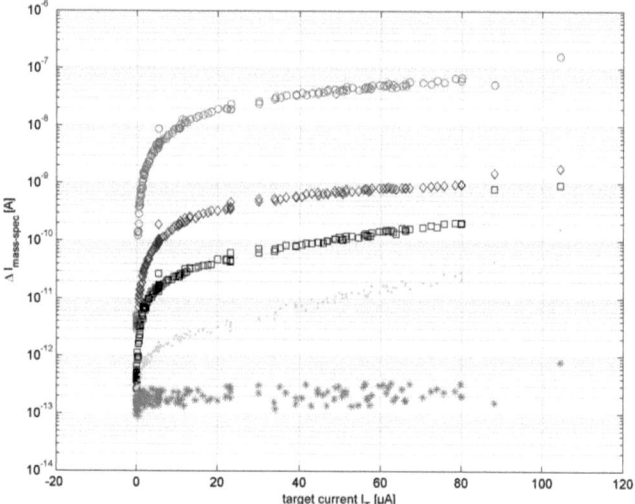

Figure 9.8.: Increase for the reduced masses m/q of the detected ions for 2(○), 14(◇), 44(□), 85( ) 87( ) where the background ion current is already corrected. The intensity in ion-current at the QMA therefore shows a proportionality $\Delta I_{mass-spec} \propto k \; I_T$, a linear increase with electrons impinging the target.

Part II.
Magnetic transport of cold atoms into a cryogenic environment

10 Concept of a magnetic transport for cold atoms

This part of my thesis describes the second approach establishing an ensemble of ultra-cold atoms in a cryogenic environment, which is yet implemented in the experiment (see section 4.3). In contrast to *part I* which describes a novel source for cold atoms, the concept of a magnetic transport for cold atoms rather relies on a combination of well established techniques, known from cold atom physics.

The concept is in principle based on a general transport scheme between two chambers, the MOT-chamber and the experimental chamber. The big advantages which come along by separating the production and preparation of ensembles of cold atoms from the real experiment in the experimental/science chamber, often overbalance drawbacks of more complicated and sophisticated setups.

The magnetic transport is among other schemes[1], only based on magnetic fields, resigning any optical components or laser light. Starting with cold atoms trapped in a magnetic gradient field (see section 5.3), the magnetic transport can be reduced to moving the trap minimum through space. This further opens two possibilities for a magnetic transport scheme to be realized, either moving magnetic coils, or a configuration of static magnetic coils, where time dependent currents generate a moving magnetic trap [175].

The first was already used at NTT[2] to transport cold atoms inside a cryogenic environment [19], while for room-temperature cold atom experiments, spacial moving magnetic coils were first demonstrated in [176], and since then repeatedly done [177].

The second possibility relies on a scheme, first realized by *Greiner et al.* [175], which was used to transport atoms just horizontally. This scheme can not exclusively be applied to the

---

[1]For the experimental concept, other transport schemes were studied, which includes the optical transport even of a BEC [174] and a transport of cold atoms into a cryogenic environment using optical tweezer [21]
[2]NTT Basic Research Laboratories, NTT Corporation, Kanagawa - Japan

10. Concept of a magnetic transport for cold atoms

experiment described within this thesis, as the cryogenic radiation shields of the experimental chamber restricts access to certain directions (see Fig.10.1, even if an intermediate radiation shield between between 4K and 300K is implemented.
In addition, a simple vertical or horizontal transport as depicted in inset a) and c), would demand trapping currents which are not reasonable achievable and therefore indicates a combination of a horizontal and a vertical transport, as depicted in inset d). As the horizontal transport has already been proven, the vertical magnetic transport scheme was developed[3] in the framework of the Quantum Interconnect Experiment at the Vienna University of Technology/Institute for atomic and subatomic physics. Parts of the novel transport scheme are also described in the Diploma thesis by N. Lippok [178].

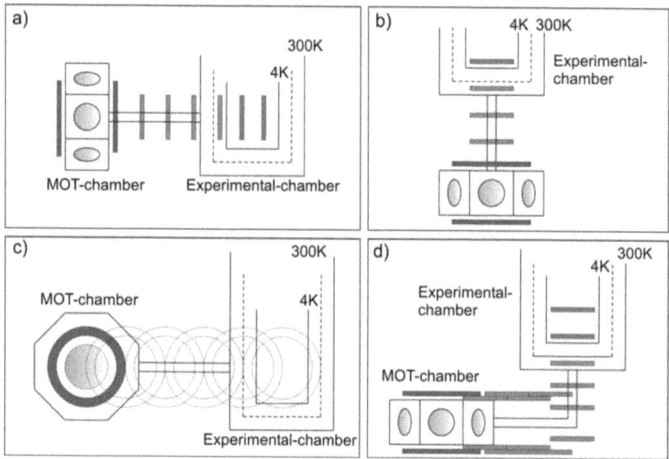

Figure 10.1.: Dierent setups for the magnetic transport into a cryogenic vacuum chamber are shown. While inset a) an c) show a concept trying to avoid a round-the-corner transport, as well as coils which fit in between the thermal radiation shield, inset b) shows a vertical transport, depicting the transport coils in orange. Even if the MOT-chamber is build very compact, unreasonable high currents in the red MOT-coils must be applied, to vertically pull out the atoms from the preparation chamber. Inset d) shows a more convenient concept, combining the well established horizontal transport with a vertical transport which allows to penetrate the thermal shield of the cryogenic vacuum chamber.

---

[3]The author mentions for completeness: It is known that also an other experiment (Thywissen Group, Toronto) tried to implement the vertical magneto-static transport, where so far no successful implementation, has been reported.

10.1. Demands for a magnetic transport line

Principles of a magnetic transport

The geometry for a horizontal moving magnetic trap must at least allow for a symmetry in two directions, whereas the third direction defines the transport axis. Referring to section 5.3, the most simple trapping geometry is an anti-Helmholtz configured magnetic quadrupole trap with a magnetic field minimum of $B = 0$ at the center. The question of how many coils participate in the trapping potential is quite crucial and strongly depends on the following demands.

Trapping gradient and size of trapped atom cloud

An anti-Helmholtz configuration provides the strongest magnetic field gradient at constant current, which is defined to be 130G/cm in the vertical direction. The size of the trapped atom cloud is furthermore not only depending on the steepness of the trap, but also on the temperature of the atoms. With maximum temperature of $200\mu K$ in the initial quadrupole trap before transport, this results in a certain width of the cloud, which can be estimated the following way.

Assuming the atoms to have a nearly Gaussian distribution, which is not completely true for a linear trap such as a quadrupole trap, but can be applied with clear conscience, the mean thermal energy can be related to the potential energy in the trap. From [176], a feasible estimation reads as

$$k_B T = \frac{2}{5}\mu_B g_F \frac{\partial B_r}{\partial r} \; FWHM \qquad (10.1)$$

where the potential energy of the atoms, defined by the Lande-factor g_F, and Bohr's magneton μ_B together with the magnetic field at the full-width half-maximum dimension $FWHM$ defines the mean top dead center of the oscillating atoms in the trap. Applying at least a radial gradient B_r of 65G/cm and a temperature of $200\mu K$ of the atoms in the trap, the full-width half-maximum turns out to be = 2.3mm.
In addition a linear magnetic trap with $B = 0$ at the trap minimum, intrinsically sets a lower limit for the temperature as Majorana-spin-flips (see section 5.4) occur, if the rate of magnetic field change is not much smaller than the Larmor-frequency of the rubidium atoms. For temperatures in the region of $200\mu K$ and a field gradient of $B = 130G/cm$ in the vertical direction, this would not cause major losses to the trapped atoms.
Finally the estimated size of the cloud implements some restrictions to the transport tube which has to account for the spacial radius of the trap, since a set of dierential pumping stages will be introduced (see section 11.2.2) to improve vacuum in the transport section.
Nevertheless the overlapping geometry of the coils prevent all coil-pairs to be in the ideal distance from each other. Furthermore to keep the trapping gradient constant over the transport distance the aspect ratio is introduced.

Aspect ratio of the trap

The aspect ratio of the trap $A = \frac{\partial B_y / \partial y}{\partial B_x / \partial x}$ must be constant, where y denotes the transport direction, and x denotes the third direction, orthogonal to the vertical axis z. A variation, or oscillation of the aspect ratio during the transport, would lead to a heating (section 5.4) of

10. Concept of a magnetic transport for cold atoms

the atomic ensemble, and must therefore be prevented. Except acceleration and deceleration of the trap, the atoms will than refer to an inertial trapping-frame which does not vary with time.
The defined aspect ratio, therefore fixes the trapping gradients for all three directions, as $\partial B_y/\partial y$ is related to constant vertical gradient of 130G/cm at each position, starting with $\frac{\partial B_z/\partial z}{\partial B_y/\partial y} = 2$ for an ideal anti-Helmholtz configured quadrupole trap at the beginning of the transport.

Current drives

The vertical gradient $\partial B_z/\partial z$, as well as A is valid for both, the horizontal and the vertical transport. Whereas in the horizontal transport the different trapping currents can be chosen to be positive only for a certain assumption of A, the geometry of the vertical transport demands bipolar currents.

Power consumption

Realizing a magnetic transport line with more than 15 coils over a distance of half a meter, the power consumption of the coils plays an important role. The strong vertical gradients of up to $130G/cm$ demand a compact design of the coils, limited by the dimensions of the transport tubes and the MOT-chamber. As the electric power scales as $P \propto I^2 \times R$, and $I \propto \partial B_z/\partial z \propto \frac{1}{r_{coil}^2}$, respectively $R \propto r_{coil}$, this results in the strong exponential relation $P \propto r_{coil}^5$ for a fixed gradient in a anti-Helmholtz configured quadrupole coil-pair. This demonstrates the strong demand for a compact design of the transport section.

10.2. Basic design considerations

Considering the demands shown above, the following additional design considerations will lead us to the setup for which the transport currents can be calculated.

The push-coil

Within the given dimensions of the MOT-chamber (see section 11.2.1) to allow optical access for laser-cooling, and the implementation of vacuum pumps, an additional coil for pushing the atoms out of the chamber is needed. It would also be advantageous if this coil is as close to the center of the chamber as possible to allow for strong fields in the order of hundreds of gauss to push the trap minimum towards the first transport coils. Even a rough estimation of this push-coil shows that this coil has to withstand currents of more than 100A. An optimal configuration would therefore be compact shape and a close as possible position to the trap center.

Water-cooling

A compact design of nearly anti-Helmholtz configured pairs of coils would not withstand the high temperatures produced by the trapping currents, without a cooling system. Nevertheless a water-cooling-system needs space to be implemented and may demand deviations from the ideal coil geometry.

Superconducting Coils

The following considerations for the superconducting coils inside the cryostat even face the drawback of an intrinsic limitation: to achieve a certain magnetic flux resulting in proper trap gradients at fixed coil geometries, either the current in the coil must be increased, or the number of windings must be cranked up, while increasing the inductance of a coil. As in a 4K environment cooling power is limited, the number of windings has to be increased dramatically, to minimize heat dissipation through the normal conducting part of the wire, or solder contacts, even if superconducting wires are implemented. The fixed vertical gradient of 130G/cm will at least demands a geometry of the coils similar to the outside room-temperature coils. Otherwise it would be difcult to keep $\partial B_z/\partial z$ and A constant within certain limitations.
It should be noticed that the number of coils in the cryogenic environment is further eected by design considerations of the 4K section (see section 12) and will be derived in a subsequent chapter.

10.3. Simulation of the magnetic transport

To calculate the transport sequence, further input aside the dimensions and positions of coils is needed. Several coils are participating in the transport simultaneously, and provide the trapping gradient at a given position. Using *Biot-Savart's* law

$$dB(r) = \frac{\mu_0}{4} I dl \frac{r - r'}{|r - r'|^3} \quad (10.2)$$

which gives the magnetic field dB at the position r of an infinitesimal wire with length dl at the position r', the magnetic field for a sum of participating currents flowing in coils with windings N can be written as

$$B(r)_{r \to x} = \frac{\mu_0}{4} \sum_{i=1}^{m} \sum_{j=1}^{N} \int I_i dl \times \frac{r - r'_{i,j}}{\left|r - r'_{i,j}\right|^3} \quad (10.3)$$

where I_i is the current running in the $i-th$ coil. Assuming the following conditions discussed in section (10.1), Eq. (10.3) builds up a set of equations, which allows for as many conditions as free parameter occur in the system.

1. $B(x) = 0$ at trapping position
2. $\partial B/\partial z = 130 G/cm$
3. $A = \frac{\partial B_y/\partial y}{\partial B_x/\partial x} = const. = 1.62$

To allow for those three conditions, the participation of 3 pairs of quadrupole coils is necessary. Except for the beginning and the end of the horizontal transport, where just the MOT-coil and the push-coil, and H_4 together with the quad-coil pair $V_1 - V_2 = H_5$ are providing the trap gradient, the conditions 1.)-3.) shown above can be fulfilled. Setting the aspect ratio to $A = 1.62$ would additionally allow to restrict the transport currents in the horizontal coils to be positive and therefore unipolar power supplies can be used.

The scheme is depicted in Fig.(10.2) where a) illustrates the transport for two participating coil-pairs. This would cause the aspect ratio to oscillate, and hence the atom cloud to

10. Concept of a magnetic transport for cold atoms

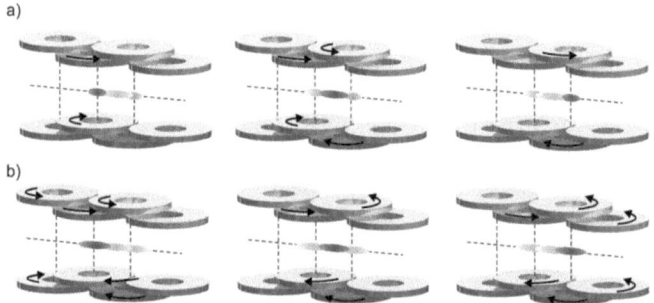

Figure 10.2.: Related to the idea from *Greiner et al.*, the trap gradients can be kept constant as depicted in the lower row, using three pairs of coils simultaneously instead of two pairs as shown in the upper row.

be heated up. The three pics in b) show how a constant aspect ratio over the transport is maintained, using 3 coils which participate in the transport.

For the vertical transport four coils should participate in the transport except the beginning and the end of the transport, allowing the following conditions for the currents.

1. $|B(z)| = 0$:at trapping position

2. $|B_z|' = 130 G/cm$

3. $|B_z|''_{z<z'} = 0$:resulting in a linear trapping gradient over a wide range

4. $\sum_{i=1}^{4} I_i = 0$

Condition 3.) assures that for an arbitrary superposition of magnetic fields, assuming a final magnetic potential defined by a polynomial of 3rd order as $B(z) = az + bz^2 + cz^3$, the trap has a linear gradient in a region defined by z'. As the general condition $B'_z = 130 G/cm$ would just demand the gradient in the trap center to be constant, $B(0)'' = 2b + 3cz$ both, the quadratic coefficient and the cubic coefficient demands to be at least for a region $z < z'$ very small. This ensures a linear trap gradient over a wide trapping range.

It turns out that at least the first 3 conditions satisfy the demands to keep the aspect ratio during the vertical transport constant, which would just mean that 3 participating coils are enough. Calculations of this setup nevertheless show, that the slope of the currents are than not steadily dierentiable. As the implementation of a fourth coil in the transport seems advantageous, the 4th condition is introduced, to maintain steady current curves.

10.3. Simulation of the magnetic transport

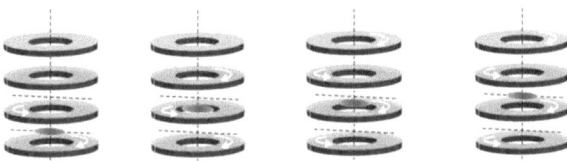

Figure 10.3.: From left) Beginning of the vertical transport sequence is shown, where in the corner the trap gradient is maintained by a quadrupole configuration with two currents on. Then the current in vertical coil 3 and 4 is stepwise increased until the atoms have moved to the equilibrium position between coil 2 and 3. This sequence is continued with four participating currents until the trap reaches the final position, where again only 2 coils participate for the magnetic trapping field

10.3.1. Coil configuration

Nevertheless the dimensions and positions of all coils participating in the magnetic transport scheme are given in Tab.(10.1) and Tab.(10.2) which refers to the setup of coils shown in Fig.(10.4).

Considering all demands for the transport scheme, the dimensions and positions of the coils can be defined, which of course for the superconducting section in the cryogenic environment also depends on some additional demands for the setup in the cryogenic section (12.2). Nevertheless, the final configuration and calculation for the transport was obtained in an iterative way, trying several setups and current-schemes. Details on these iterations can be found in [178].

The setup of the coils is shown in Fig.(10.4), depicting the dierent coils at room temperature, as well as showing the dierent size for the superconducting vertical transport coils.

Tab.(10.1) gives the configuration for the horizontal transport, where the last horizontal quad-coil pair $H_5 \Longleftrightarrow V_1 - V_2$ also participates in the vertical transport. For the winding-scheme of the conical push-coil ($N_{total} \approx 107$), refer to the proposed winding-scheme shown in Appendix (G.1) and to [178] for the characterization of the magnetic field.

Tab.(10.2) shows the configuration for the vertical part of the transport where $V_1 - V_5$ are coils operating at room-temperature, and $V_6 - V_9$ are located in the cryogenic environment. The inductance L' for the superconducting coils is given for $f = 20 Hz$. The DC-value is therefore slightly higher. R_{300} of the sc-vertical transport coils is measured at lab-temperature and even changes within several %[4] and usually is around 2100 ± 70. The push coil horizontal distance is measured from the nearest point of the coil to the vertical MOT-axis. For a detail winding scheme of the push-coil see Appendix (G.1).

10.3.2. Calculations of the current

To improve accuracy, the finite dimensions of the coils are used for the calculation of the currents, considering all windings as written in Eq.(10.3).

---

[4]It depends on environmental influences, whether the coils are already implemented or were just lying around

10. Concept of a magnetic transport for cold atoms

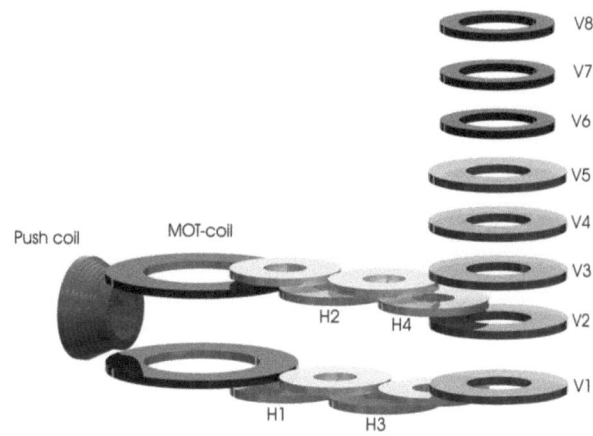

Figure 10.4.: Coil setup for the magnetic transport with the conical push-coil on the left, the MOT-coils which also establish the initial magnetic trap at the beginning of the transport, the yellow horizontal pairs of transport coils and five vertical, normal conducting transport coils, colored orange. The last transport coils are superconducting and are depicted in blue.

coil	push-coil	MOT-coil	H_1	H_2	H_3	H_4
$R_i[mm]$	34	36.2	13.3	13.3	13.3	13.3
$R_o[mm]$	52...72 (conical)	59.5	34.3	34.3	34.3	34.3
$h[mm]$	30.6	5.1	5.1	5.1	5.1	5.1
$d[mm]$	0	55.1	71.5	55.1	71.5	55.1
$\Delta x[mm]$	-56.6	0	60.5	95	129.5	137
N_{axial}	12	2	2	2	2	2
N_{radial}	6-13 (G.1)	23	23	23	23	23
$R[\ ]$	0.102	0.023	0.0126	0.0126	0.0126	0.0126

Table 10.1.: Coils participating in the horizontal transport are shown, except H_5, which also denotes as V_1 and V_2 and is given in the corresponding table for the vertical transport: R_i, R_o are inner and outer radius of the coil, h the hight, d the mean distance between the pair of coils, Δx the horizontal position in respect to the vertical MOT-axis and N_{axial} and N_{radial} the number of windings in the corresponding directions. The ohmic resistance R gives the serial ohmic resistance of a single coil. For H1-H4 and the MOT-coil, the resistance must be doubled.

Horizontal transport

Solving the set of equations defined by Eq.(10.3) with the certain conditions, the currents for the horizontal transport in dependence of the transport position x can be calculated. As

10.3. Simulation of the magnetic transport

coil	V_1	V_2	V_3	V_4	V_5	V_6	V_7	V_8	V_9
$R_i[mm]$	20	20	20	20	20	24	24	24	24
$R_o[mm]$	43	43	43	43	43	35	35	35	35
$h[mm]$	5.1	5.1	5.1	5.1	5.1	5	5	5	5
$\Delta z[mm]$	-20	20	50	80	110	140	170	200	230
$\Delta x[mm]$	210.4	210.4	210.4	210.4	210.4	210.4	210.4	210.4	210.4
N_{axial}	2	2	2	2	2	36	36	36	36
N_{radial}	20	20	20	20	20	84	84	84	84
$R_{300}[]$	0.0115	0.0115	0.0115	0.0115	0.0115	≈ 2100	≈ 2100	≈ 2100	≈ 2100
$R_5[]$	—	—	—	—	—	≈ 0.55	≈ 0.55	≈ 0.55	≈ 0.55
$L'[mH]$	—	—	—	—	—	470	582	583	—

Table 10.2.: Coils participating in the vertical transport: R_i, R_o are inner and outer radius of the coil, h the hight, Δz the vertical position of the coil center, Δx the position to the vertical MOT-axis, N_{axial} and N_{radial} the number of windings in the corresponding directions. R_{300} gives the ohmic resistance at room-temperature, R_5 at $T = 5K$ and L the DC-inductance of the coil.

mentioned in section (10.1), at the beginning of the transport sequence only the MOT coil is on, demanding an initial current I_{trap} which generates the trap gradient of $130G/cm$ in vertical direction. In addition, the aspect ratio due to symmetry reasons equals $A = 1$.
As the position of the trap and the trapping gradient $\partial B_z/\partial z$ in vertical direction are the strongest arguments, the aspect ratio in the beginning of the transport varies as the push coil and the horizontal coil H_1 are switched on, and gets constant when at least the MOT-coil, H1 and H2 are on.
Geometric restrictions define the distance between V_1 and V_2 which acts also as the last horizontal trapping coil-pair at $x = 210.4$. For consistency the trap gradient in vertical direction in the beginning of the vertical transport has to be $130G/cm$, which is also valid for the end of the horizontal transport. As therefore a quadrupole trap just defined by the currents in V_1 and V_2 has to be established, the aspect ratio for the end of the horizontal transport can not be kept constant and changes from A=1.62 back to A=1 as shown in Fig.(10.5). Due to the short and uniquely change of the aspect ratio we expect that this will not dramatically effect the temperature of the trapped atoms, regarding to first characterizations of the horizontal transport line [178].
The calculated currents for the horizontal transport are given in Fig.(10.6), showing the tremendous high current for the push coil, even for the compact setting of this coil. It occurs that the currents are not completely smooth over the complete transport distance, nevertheless the curves exhibit a continuous current-shape without any rapid changes.

Vertical transport

In contrast to the horizontal transport, which allowed for positive transport currents, setting $A = 1.62$, the intrinsic configuration of the vertical transport demands bipolar currents. In addition the aspect ratio of $A = \frac{\partial B_y/\partial y}{\partial B_x/\partial x}$ is defined along the transport direction x, which makes no sense for the vertical transport along the z-direction, as the symmetry of the coils would always result in $A = 1$.
Together with the second vertical transport condition $B'_z = 130G/cm$, constant gradients

10. Concept of a magnetic transport for cold atoms

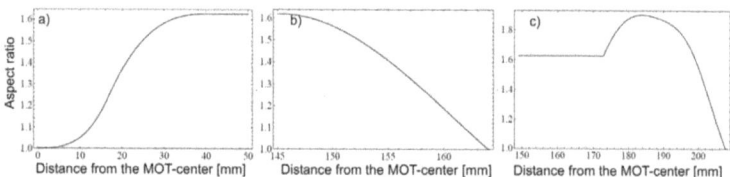

Figure 10.5.: Aspect ratio of the magnetic transport at the beginning and the end of the scheme. Inset a) shows $A = 1 \to A = 1.62$ at the beginning of the horizontal transport. Inset b) and c) depicts how at the end of the horizontal transport the aspect ratio decreases to $A = 1$ as the atoms are trapped at the final position x_{final} in the center of H_4, inset b) and for $x_{final} = 210.4$, in the trap at the corner of the transport, maintained by H_5, inset c). The graphs are taken from [178].

in all three direction are maintained as $B'_x + B'_y + B'_y = 0$ is fulfilled, resulting in $B'_x = B'_y = 65 G/cm$, which is obvious due to the radial symmetric geometry and therefore $C = \frac{\partial B_z/\partial z}{\partial B_y/\partial y}$, stays constant. The aspect ratios C_0 and C_{final} are therefore as well as during the whole transport two times as big as the radial aspect ratio A.

As the distance $d = 40mm$ between the vertical coils V_1 and V_2, is much bigger than between all other vertical coils, much more current is demanded in the beginning of the vertical transport, as shown in Fig.(10.7).

The vertical currents are depicted in Fig.(10.7), showing different currents in the coils for the room-temperature and the cryogenic environment setup. During the complete vertical transfer, four coils are participating in the transport, except the beginning and the end of the sequence. Even if the currents change at the intersection of the normal- and the superconducting part of the transfer, the ampere turns $I_i \times N_i$ stays roughly constant, as of course the coil geometry changes slightly.

Transport parameter

The calculation of the currents ends with the position dependent solution for the magnetic fields, and therefore in $I_i = f(x)$. Nevertheless the experimental control just refers to time-dependent values for the control signals, and therefore the resulting currents must be transformed $I_i = f(x) \to I_i = f(t)$. This transformation is done different for the horizontal and the vertical part.

Hence, defining parameter are transportation time T for the horizontal part, and the time after which the maximum velocity is reached T_{max}. As the horizontal transport distance[5] can be changed, or is by definition limited to $210.4mm$, the particular velocity and acceleration can be calculated. Therefore the acceleration is set to a maximum at $T_{max}/2$ following a polynomial function as

$$a_{hor-1}(t) = c_0 + c_1\ t + c_2\ t^2 + c_3\ t^3 \qquad (10.4)$$

for $0 < t < T_{max}$, with $a_{hor} = 0$ at $t = 0$, $t = T_{max}$ and $t = T$ and further boundary conditions such as $v_{hor} = 0$ at $t = 0$ and $t = T$, with $v_{hor}|_{t=T_{max}} = v_{max}$, and symmetric

---
[5]For test-reasons a transport scheme can be run, with forth and back transport of the atomic cloud.

10.3. Simulation of the magnetic transport

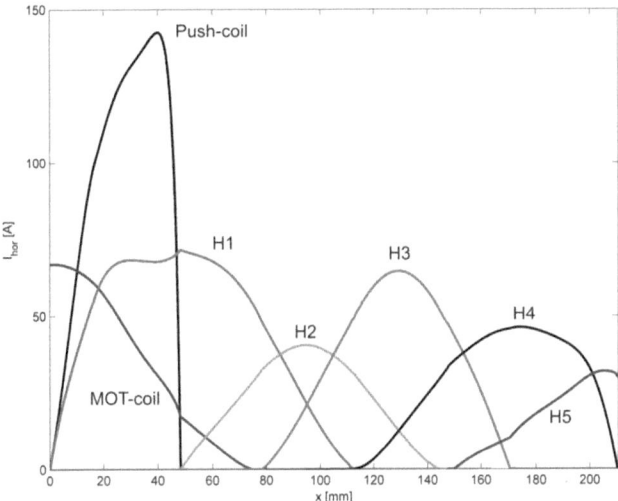

Figure 10.6.: Currents for the horizontal transport are shown. Beginning with the push-coil, the MOT-coil and the horizontal coil H_1, the trap is shifted to $x \approx 50mm$. From then on, the trap is maintained by the MOT-coil pair, H_1 and H_2. After $x = 80mm$, the transport works with the horizontal coils only. As the aspect ratio is set to A=1.62, all participating currents are positive without the need for bipolar current sources.

10. Concept of a magnetic transport for cold atoms

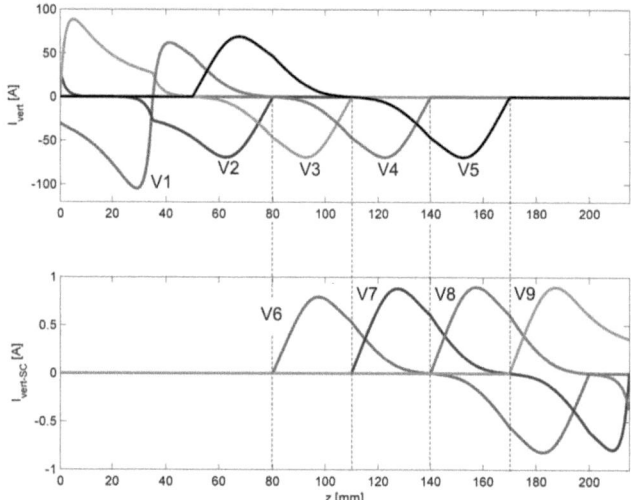

Figure 10.7.: Vertical transport currents for all nine vertical transport coils are presented. Starting with the normal conducting quadrupole trap, maintained by V_1 and V_2, the third coil is switched on, almost reaching 100A. After a vertical transport distance of $z = 50mm$, four coils are participating, and after $z = 80mm$ the same current shape follows in a repetitive way, while at this position the first superconducting coil already participates. Notice therefore that the current in V_6 is slightly smaller than in V_7 and V_8, as the distance between V_5 and V_6 was found to be not accurately $30mm$. Up to $z \approx 200mm$ four coils maintains the transport while the transport ends with $I_7 \to 0$ and I_8 and I_9 at the same value, just driven in opposite direction through the coil. In subsequent chapters, the positive half-wave of the current will be indicated with $'a'$, while the negative will exhibit $'b'$, and therefore $I_6 \to I_6 a$, $I_6 b$.

10.3. Simulation of the magnetic transport

conditions for the deceleration during $T_{max} < t < T$. The velocity $v(t)$ and distance $s(t)$ therefore origins from the integral of $a_{hor}(t)$ as shown for the vertical transport in Eq.(10.5).

For the vertical part, again the transport distance is the defining parameter, whereas the maximum velocity v_{vert} and acceleration a_{vert} can be chosen independently[6]. The vertical transport time therefore results from the integral of the velocity, given as

$$\frac{z_{max}}{2} = \int_0^{T/2} v_1(t)dt \quad \text{with:} \begin{cases} v_1(t) = \int_0^{T/2} a_1(t)dt \\ a_1(t) = c_{10} + c_{11}\ t + c_{12}\ t^2 + c_{13}\ t^3 \\ a'_1(t) = c_{11} + 2c_{12}\ t + 3c_{13}\ t^2 = 0 \quad \text{for t=0, t=T/2} \end{cases} \quad (10.5)$$

and similar for $v_2(t)$ and $a_2(T)$ if $T/2 < t < T_{vert}$.

---

[6]They indeed can be independently chosen from the maximum transport time, but are strongly limited by the switching times of the superconducting transport coils (see section 12.2.7)

11 Setup of a magnetic transport line at room temperature

The room-temperature transport line contains the setup of the MOT-chamber, with the push-coil, the horizontal transport coils $H_1 - H_4$, and the vertical coils $V_1 - V_5$. Additional the laser-system provides the light for cooling, optical pumping and imaging of the atoms, whereas compensation coils cancel the earth magnetic field to adjust and optimize polarization gradient cooling and optical pumping. Together with the water-cooling system applied at all transport coils, a redundant temperature control system allows to handle the high currents in the transport coils. To reduce the number of power supplies participating in the transport, the transport currents are distributed to the coils via MOSFET-benches and H-bridges (see Appendix B). Details on the vacuum system setup can also be found in [178], whereas parts of the laser-system setup are described in [154].

11.1. Laser system

To trap and cool ^{87}Rb in a MOT and further in an optical molasses (see section 5.3), a laser system is need which provides a line width smaller than the natural lifetime of the proper Rb-transition. The light for cooling, optical pumping and imaging, is therefore at $= 780.246nm$ as the $5^2S_{1/2} \to 5^2P_{3/2}$ D2-line of rubidium provides a closed transition for laser cooling with a cooler laser at $5^2S_{1/2}|F=2\rangle \to 5^2P_{3/2}|F'=3\rangle$ and a repumping transition at $5^2S_{1/2}|F=1\rangle \to 5^2P_{3/2}|F=2\rangle$.

The following section is divided into four parts describing briefly the Locking techniques used to stabilize the frequency, showing the setup of the different lasers and the optical setup for preparing the experimental laser-sequence.

Since semiconductor lasers are fully tunable, reliable and available in a wide frequency range, those lasers are a good choice to provide the light suitable for the Rb-hyperfine transitions. Hence, attention has to be paid on the reduction of the line width, the temperature dependence and the stability due to mechanical vibrations. Exhausting information and a general overview on the stabilization and line width reduction on semi-conductor laser-diodes

11. Setup of a magnetic transport line at room temperature

can be found in related work [154].

11.1.1. Locking Techniques

Laser cooling by means of Doppler- and sub-Doppler cooling, demands frequency stabilized laser, in order to fall below the natural line width of ^{87}Rb. Using conventional semi-conductor laser diodes with a line width in the order of dozens of MHz, this can be achieved by external gratings or cavities, as applied in our research group [179] and by an active electronic feedback of the laser frequency to reduce the influence of mechanical vibrations, oscillations of the temperature and technical fluctuations, which either lead to deterministic or random variations of the laser frequency.

The principle behind the locking techniques is to compare the output laser frequency with a reference frequency such as an atomic transition, or other stabilized lasers, in order to lock the output frequency to such a signal. The common properties of an error signal must carry the information about the deviation and the direction of the deviation from the current output frequency in respect to the reference frequency.

Ideally error signals can directly influence the laser current, which allows for a fast controlling of the frequency, as it is commonly done for laser diodes. In addition, the length of an external resonator, used for line width reduction can be changed, in order to access the frequency.

Fig.(11.1) is taken from [179] and shows both, the D2-line of ^{87}Rb, inset a), with the transitions for the cooling and the repumping light needed for a MOT, as well as the transition for optical pumping and imaging. Inset b) therefore shows a Doppler-free absorption spectroscopy of the D2-line with the hyperfine transitions $5^2S_{1/2}|F=2,m_F\rangle \rightarrow 5^2P_{3/2}|F',m'_F\rangle$ and the cross-overs which are used to lock the lasers. Following the next paragraphs, the locking techniques are described which are used to stabilize the lasers.

Laser locking with Frequency Modulation Spectroscopy

Doppler-free saturated absorption spectroscopy is used to implement a frequency modulation (FM) lock. A glass cell, containing ^{87}Rb vapor is pumped with a strong beam and probed with a weak, phase modulated probe- beam in order to maintain a Doppler-free spectrum of the hyperfine transitions. Symmetric sidebands around the carrier frequency occur on the signal due to the phase modulation, and a beat signal of the sidebands with the carrier frequency can be recorded e.g. on a photodiode. As the beat signal is frequency dependent, and is zero at resonance frequency, it is well suited to act as an error signal which can be used as a feedback on the laser-frequency.

It should be noticed that the FM-lock can be established in two dierent frequency regimes, modulating the carrier frequency either with low or high frequencies in the RF-regime from kH-MHz. A FM-lock using low modulation frequencies is usually denoted as wavelength-modulation (WM), whereas high modulation frequencies confuses with the miss leading denotation as frequency-modulation spectroscopy, as commonly the use of both, high and low modulation is denoted as frequency-modulation (FM).

Frequency o set lock

Frequency oset (FO) locking schemes provide in contrast to the FM-locking technique the possibility to not only lock the laser to a certain frequency, but also allows for tuning the

11.1. Laser system

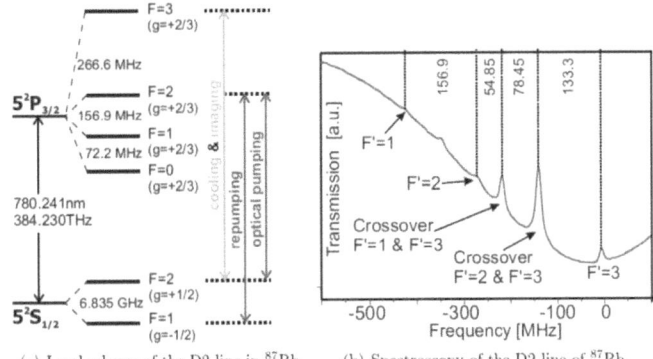

(a) Level scheme of the D2 line in ^{87}Rb.   (b) Spectroscopy of the D2 line of ^{87}Rb.

Figure 11.1.: Inset a) shows the D2-line of ^{87}Rb with the cooling, repumping, imaging and optical pumping transitions. A Doppeler-free saturation spectroscopy applied to the D2-line is shown in inset b) with the hyperfine transitions and the cross-overs which are used to lock the lasers. The figures are taken from [179].

frequency in a wide range. Of course this can also be done by AOMs or EOMs, but always with the drawback of loosing light intensity. As the denotation already implies, with the use of a FO-lock [180], an offset frequency can be applied, whereas the frequency difference is kept constant. Therefore a reference laser is needed which is stabilized and locked to a certain reference frequency. Beating the reference laser with a second, so called master laser, an error signal from the beating signal can be generated, which allows for manipulation via a voltage controlled oscillator. A principal setup of such a frequency offset lock is given in Fig.(11.2). Light from the reference- and the master-laser are superimposed on a 50:50 beam splitter and detected with a fast avalanche photodiode.

The beating intensity can be written as

$$I \propto cos\left(2\,[\Delta\,t - \frac{d}{dt}\Delta\,(t)]\right) \quad (11.1)$$

considering the difference frequencies of the master and the reference laser $\Delta = \,_1 - \,_2$, and the time dependent phase fluctuations.

The beat note signal is further mixed with a signal from a voltage controlled oscillator (VCO) and after a low-pass filter split into equal parts. Implementing a time delay and further mixing an error signal is processed which, at every zero-crossing of the cosine, depends only on the VCO frequency.

11. Setup of a magnetic transport line at room temperature

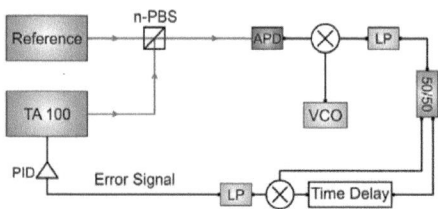

Figure 11.2.: Locking-scheme of the frequency offset (FO) lock

Figure 11.3: *Piere Luigi,* the reference-laser with its FM locking scheme is shown in the upper left corner. The laser beam is overlaid with a small part, outcoupled from the TA-maser in the NPBS. The TA-master itself is locked via FM lock, via an error signal processed as shown in Fig.(11.2). The signal detected by the fast APD at the bottom of the figure, processes the beating signal to create an error-signal at which the TA-master is locked. The fiber coupler in the upper right, collects the amplified high-power laser light and guides it to the optical setup on the table.

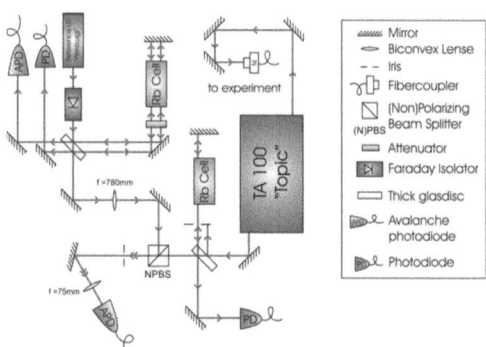

11.1.2. Cooler-Laser and amplifier

Both, having a huge cooling-laser power, while providing at the same time full frequency tunability over a range of more than 100MHz is suitable for the experimental setup. To achieve this, the cooler-laser was chosen to be realized within a FO-lock scheme, locking the master laser of a tapered amplifier to a reference laser. The error signal is maintained from a beat-node signal of the superimposed laser beams. Depending on the frequency range of the voltage controlled oscillator implemented in the FO-lock scheme, and the AOM implemented in the cooler-line, the TA-master frequency can nearly arbitrarily chosen between -70MHz and +70MHz around the resonance of the cooling transition. This now allows to use the cooler-laser for the cooling-transition in the MOT, for polarization gradient cooling in the molasses sequence, for imaging, and far detuned in-situ imaging at Zeeman-shifted frequencies up to +70MHz just by controlling the VCO-frequency.

The reference laser

The reference laser, also entitled *Piere Luigi*, is a FM-locked diode laser which was built by ourselves [154]. The laser diode setup is extended by an external cavity, to reduce the linewidth, where an RF-modulation of 20MHz is applied to the laser diode current. In addition to the work in the thesis of R. Amsüss, the setup of the laser-locking has been improved, by replacing the WM-lock by a FM-lock to increase the stability of the lock. As *Piere-Luigi* acts as a reference laser for the FO-lock, laser power is not crucial.
The laser is locked to the $5^2S_{1/2}|F=2\rangle \rightarrow 5^2P_{3/2}|COF'=1,3\rangle$ transition, given in Fig.(11.1), as this is a good starting point to shift the frequency proper, using the beam for cooling, imaging and optical pumping.

Fig.(11.3) shows the setup of *Piere Luigi* in the upper left corner. The laser light passes an Faraday isolator to prevent back reflections to the laser diode. The thick glass plate provides two beams passing the Rb-vapor filled glass cell. Both beams are detected by a photodiode to view the Doppler-free saturated spectroscopy signal, and a fast avalanche photodiode for deducing the error-signal. Most of the power ($\approx 15mW$) goes through the glass plate and is superimposed with light from *Topic*, the master of the TA-system.

Cooler master and tapered Amplifier

The high laser power for the cooling transition in the MOT is provided by a commercial tapered amplifier system[1]. This system contains a master laser and an amplifier which provides the high power up to 1000mW.
The cooler master is locked via a FO-lock as described above to the reference laser, operating at the same transition as *Piere-Luigi*, $5^2S_{1/2}|F=2\rangle \rightarrow 5^2P_{3/2}|COF'=1,3\rangle$. Therefore tuning the VCO frequency allows to shift the frequency for the cooling beam nearly arbitrarily by $\approx 140MHz$, resulting in a detuning $= -70... + 70MHz$ for the cooling transition. Only the frequency rate at which the VCO can be shifted is limited to $\approx 30MHz/ms$ to prevent the laser from falling out of the lock.
For the FO-lock, a part of the master power is branched o after an optical isolator, where the remaining light is amplified without a change in coherence properties. The beam profile after the amplifier is quenched in a way, that the implementation of an optical fiber remarkable overcomes any drawback in loss of power, to maintain a nice Gaussian profile of the beam.

Fig.(11.3) shows the TA-system on the right, where $\approx 1000mW$ from the TA-amplifier chip are directed onto a fiber coupler. A small part of the master is branched o for the FO-lock. The Doppler-free saturated spectroscopy signal which helps to find the right locking point of the beat-note signal, is again maintained by coupling out a small fraction of the light to pass a Rb-vapor filled glas cell. The main part of the branched o master light is superimposed in a NPBS cube with the reference laser to deduce the error signal regarding to the scheme in Fig.(11.2).

---

[1]TA100 system from Toptica Photonics, Germany

11. Setup of a magnetic transport line at room temperature

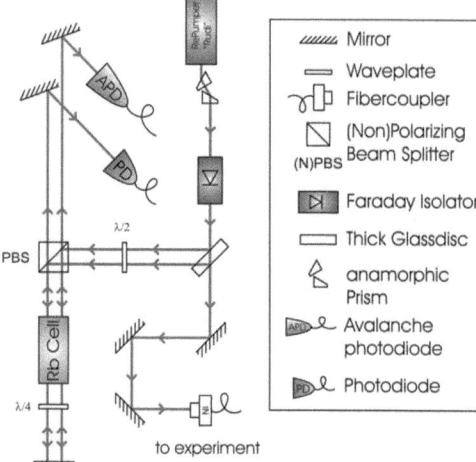

Figure 11.4: The repumper beam exits from the DFB-diode and is shaped, passing an anamorphotic prism. The thick glass-plate provides two beams which can be adjusted with a half-wave-plate. Therefore the intensity for the spectroscopy can be adjusted for the beam which ends up for the error-signal processing. This beam is detected by a fast avalanche photo diode after passing the vapor-cell while the other is just used to watch the spectroscopy signal via a simple photodiode. As in all other laser-setups shown in this thesis, the repumping light is coupled into an optical fiber to be guided to the experiment.

11.1.3. Repumper-Laser

For the repumping transition $5^2S_{1/2}|F=1\rangle \to 5^2P_{3/2}|F'=2\rangle$, a distributed feedback (DFB) laser diode is used, as it is easy to handle. It combines in a smart way the advantages of conventional laser diodes such as tunability and spectral properties, while providing a less complex setup at the same time, as no external cavity is used to reduce the line width. I use a single mode DFB diode, with a line width in the order of some MHz, designed for rubidium spectroscopy and laser cooling [181].

Nevertheless the beam profile from the DFB laser diode shows room for improvement. The laser beam passes a pair of anamorphotic prisms to improve the profile. Thereby the profile becomes more circular and can now pass the Faraday isolator. A thick glass plate couples two beams out, one each for the error signal, and one just for observation. These two beams are after passing the vapor-cell recorded with a photodiode, respectively an fast APD from which the error-signal is deduced as shown in Fig.(11.4). In order to adjust intensities a PBS is introduced which allows for adjusting the intensity via rotation of the polarization. Similar to the reference laser, the repumper is stabilized with a FM-lock, modulating the laser diode current with 20.5 MHz. Thereby the laser is locked to the crossover transition $5^2S_{1/2}|F=1\rangle \to 5^2P_{3/2}|COF'=(1,2)\rangle$ and later shifted to the repumping transition by the use of an AOM.

11.1.4. Light conditioning

To prepare the laser light for the experimental sequence using shutters and switches, to adjust/shift the frequencies of the laser beams by use of AOMs, and to shape the beam profile or changing the polarization of the beams, a comprehensive setup of optical instruments is

11.1. Laser system

needed.
Two single mode optical fibers provide the light from the separated laser-box, transferring $\approx 500mW$ from the TA100 system and $25mW$ from the DFB-laser diode to the optical table.

As the repumper laser *Rudi* is just used for the $5^2S_{1/2}|F=1\rangle \rightarrow 5^2P_{3/2}|F'=2\rangle$ repumping transition to close the cooling cycle, the frequency is shifted by a fixed value (78.5MHz) from the locking point to the accurate transition. Light from the repumper is overlaid in a PBS with the cooler light for the MOT, where a shutter between the PBS and the AOM allows for switching the light mechanically in addition to switching on/o the AOM or regulating the power level of the AOM. This three-fold mechanism ensures that there is no repumping light in the chamber when it is not supposed to be.

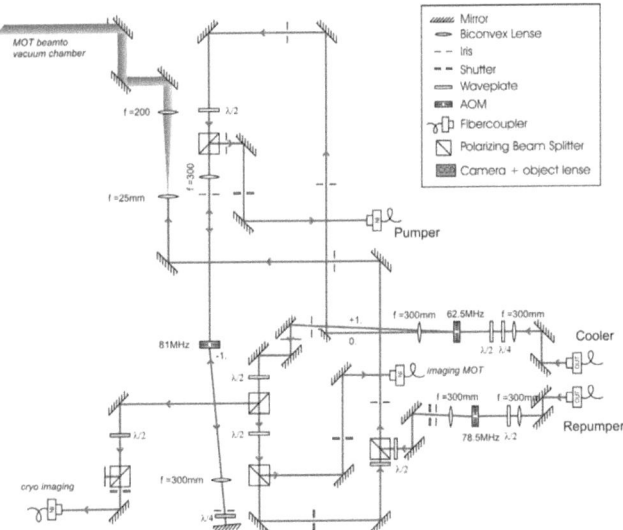

Figure 11.5.: Laserpark of the experiment which prepares the light for cooling, repumping, imaging and optical pumping

Cooler-light from *Topic* is focused onto the cooler-AOM operating at a fixed frequency of 62.5MHz, also see Fig.(11.5). To assure a high e ciency of the AOM close to 70%, the polarization is optimized using a quarter- and a half-wave plate to adjust the polarization afte the fiber. For the cooling transition and the imaging transition the first dirrected order (+1.) is used. This provides the advantage of using the AOM as a fast switch to enable the laser in the imaging sequence, as the cooling light is also used for imaging. After passing the iris, fading out all other orders (-2. -1. 0. +2.), the beam is directed onto two following PBS in order to couple out a small fraction of light for the two imaging beams, one for the lower

105

11. Setup of a magnetic transport line at room temperature

MOT-chamber, and one for the cryogenic chamber. Furtheron the beam is overlayed with the repumper light and directed onto a telescope, expanding the beam to one inch diameter. As the two lenses are not perfectly well suited for the initial beam diameter, 20% of the cooler laser power is wasted, ending with 130mW in the MOT beam. This beam is further divided into several beams and guided to the vacuum chamber to establish the MOT beams.

For pumping the atoms optically into the $|F = 2, m_F = 2\rangle$ low field seeking state, the zero order of the cooling light is used. The AOM in the double pass line as shown in Fig.(11.5), operated at $2 \times -81MHz$ therefore allows to shift the laser frequency to the pumping transition without the need of shifting the VCO frequency of the FO-lock. This results in a nice feature: If the cooler beam in the optical molasses is operated at $= -50MHz$, red detuned from resonance using the +1. order of the cooler AOM, with the double pass operated at $2 \times -81MHz$, the light can immediately ($< 0.5ms$) be switched from molasses to pumping detuning (by switching o the cooler AOM and using the 0. order for optical pumping), as the VCO frequency of the beat-lock must not be shifted.

Therefore the drawback of the double-pass AOM-line, reducing the AOM e ciency [2] is accepted, as just a small fraction of the high-power cooler light is used for optical pumping.

In conclusion, all AOM are operated at a fixed frequency. The dierent frequencies for pumping, imaging and cooling are achieved by tuning the VCO of the FO-lock between *Piere Luigi* and *Topic*. The use of either the 0. order or 1. order of the cooler AOM in addition allows to shift the frequency -54.85 MHz or +211.8MHz in respect to the locking point at $5^2S_{1/2}|F = 2\rangle \rightarrow 5^2P_{3/2}|COF' = (1,3)\rangle$ without changing an AOM frequency.

Light is switched on/o by a threefold mechanism using shutters, TTL signals to enable the AOM and via adjusting the power of the AOMs via the experimental control system (Appendix C).

11.2. Setup of the lower transport-chamber

The vacuum-chamber for the MOT and starting point of the magnetic transfer is nearly a copy of the one used in *Greiner* et al. [175], even with slight modifications. It allows optical access through 10 CF40 flanges, whereas some are also used to attach pumps and the transport tube. It was chosen to built the chamber from stainless steel, fabricated from one metal block, taking care that every flange, valve or the chamber itself is as less magnetic as possible.

11.2.1. The vacuum-chamber

The MOT-chamber setup can be seen in Fig.(11.6), where the attachments on the octagon are shown. At flange *F1*, inset b), a CF40 4-way cross is mounted which provides a full metal valve towards the turbo-molecular pump and sustains a CF40 flange with four AMDs [123] rubidium dispensers in parallel. The optical access is maintained via a CF40 window in the horizontal direction of the CF40 cross. At flange *F2*, a CF63 cross is mounted which provides the 50l/s Falcon Ion pump and a vacuum gauge, while maintaining the optical access via a

---

[2]AOMs are operated at the highest e ciency if the incoming light is polarized to the vertical axis of the AOM. Since in the double pass line the light enters and exits via a PBS, the polarization has to be turned by 90°, resulting in either a vertical polarization before or after the AOM

11.2. Setup of the lower transport-chamber

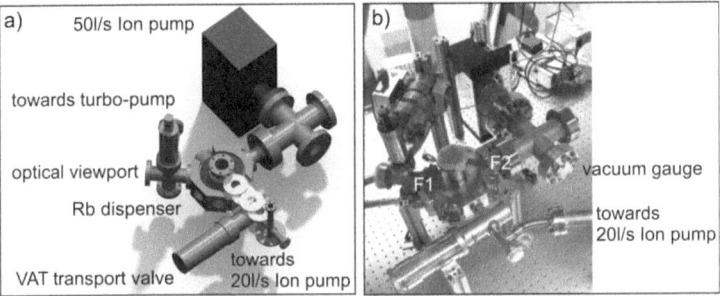

Figure 11.6.: This figure shows both, a picture and a drawing of the vacuum chamber setup for the lower magnetic trap, with the central MOT-chamber, the horizontal transport section, ion pumps and the implemented UHV valves.

CF63 window. As the turbo-pump was even initially used to pump the chamber, the ion pump allows to operate the chamber at pressures lower than $2.0 \times 10^{-10} mbar$ which is fairly enough for a MOT and a magnetic trap.
As depicted in Fig.11.6, inset a) the push-coil(green) sits as close as possible to the trap center (vertical axis of the MOT chamber). Additional clearance in the chamber grab the water-cooled rails on which the horizontal coils and the MOT coils are implemented. Inset b) shows the 20l/s ion pump attached on a different CF16 cross in the corner of the horizontal transport which was mounted for test reasons. The horizontal transport coils, together with the MOT-coil is depicted in the left inset, whereas just the vertical coil V_1 is shown.

11.2.2. The transport line

The transport tube consists of a CF16 tube connected directly between the MOT-chamber and the VAT valve [3]. Instead of the two copper gasket rings on both ends, copper tubes are built in (Fig.11.7), which acts as a differential pumping stage between the MOT-chamber and the transport section. For the corner-section of the transport two different setups where used as shown in Fig.(11.6), once just for the testing of the horizontal transport tests. As now the transport and MOT-chamber is attached to the cryogenic vacuum-chamber the setup is realized as depicted in Fig.(11.7). An asymmetric CF16 4-way cross is attached to the VAT-valve, with a 20l/s ion pump in extension of the transport tube, and a CF16 window on the bottom. The upper flange is directly attached to the cryogenic vacuum chamber (see section 13.1).

11.2.3. MOT-optics

Implementing the MOT-laser beams with attachable fiber coupler and quarter-wave plates would be the most robust way to bring the light into the chamber. In this setup the focus was

---
[3]VAT 48124-CE01 from Vakuumventile AG, CH-9469 Haag

11. Setup of a magnetic transport line at room temperature

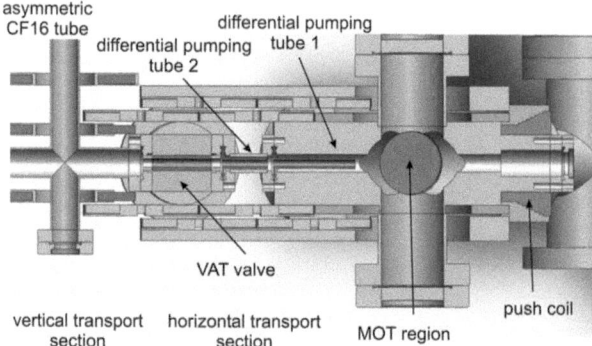

Figure 11.7.: Setup of the horizontal transport line, with the two Cu-tubes restricting the size of the trapped and transported cloud of atoms. This two tubes act as a dierential pumping stage as the pressure in the lower chamber should not limit the achievable pressure in vertical transport section or the cryogenic section.

on a compact design, which causes some challenges for the MOT optics. In a first approach a three-beam-retro-refleted MOT was achieved, simplifying the partitioning of the MOT-beams and competing with the compact design especially at the window next to the VAT-valve. As it turned out that the MOT and optical molasses could be better adjusted the setup was turned into a six-beam MOT as depicted in Fig.(11.8).

The pumping and imaging beams are guided through a separate, large CF63 window. Optical components are mounted on a separate bread board in front of the CF63 cross. Light from both fiber out-coupler is overlaid in a PBS with adjustable polarizations in front, maintained by two half wave plates and a quarter wave plate in front of the window as shown in Fig.(11.8).

11.3. Coil temperature control system

In order to prevent overheating of the coils, a threefold temperature control system was implemented, which at least allows for a redundant temperature control.
The first protection origins from a power switch o, triggered by a set of bi-metal temperature sensors[4] cutting the main power line for the main transport power supplies.
The second protection is triggered by a temperature control box using micro-controller based temperature sensors Fig.(11.9) inset b). The maximum temperature can therefore be set in two independent measuring circuits, one for the horizontal transport, and one for the vertical transport. The micro-controller based sensors are for redundancy placed next to

---
[4]The resistance of the bi-metal temperature sensors almost drops to zero, if $\approx 75°$ are exceeded. Therefore the resulting voltage drop at a two serial resistances, resulting in a control signal for a relay. The bi-metal sensors are mounted at push-coil, upper and lower horizontal rail and V1-V5

11.3. Coil temperature control system

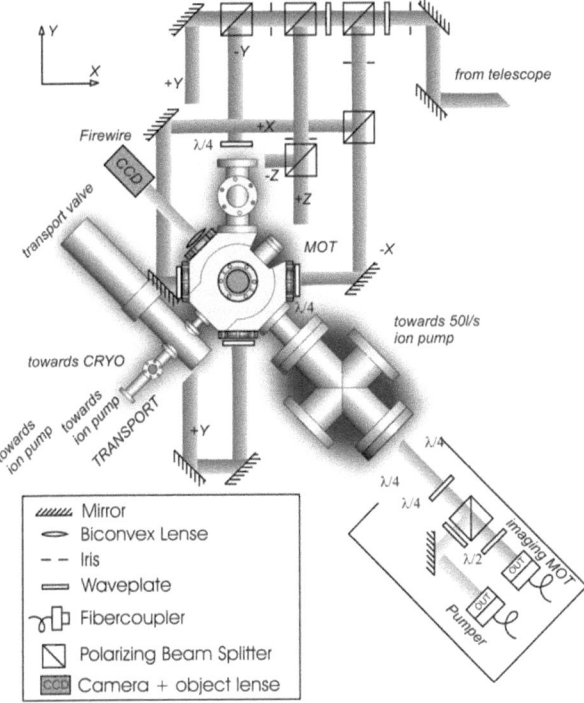

Figure 11.8.: Setup of the MOT-optics with 6 counter propagating beams from three directions guided through the chamber. To adjust the circular polarization, a quarter wave-plate is mounted in front of the vacuum-windows. The optics for optical pumping and imaging is placed on a separate breadboard in depicted in blue

the bi-metal sensors at vertical coils V1-V5, the push-coil and both sides of the horizontal rail-track.

The third protection is implemented in the switch-boxes (Appendix B) which distributes the current from the power supplies over the dierent coils. In those switch-boxes, a control-board is implemented, which measures the temperature of the copper rails, charged with up to 150A. Both, the temperature control box and the control-circuits in the switch-boxes disable the power supplies, where the temperature control box acts on all supplies, and the temperature control circuit in the failed box just aects the corresponding power supplies feeding this switch-box.

109

11. Setup of a magnetic transport line at room temperature

 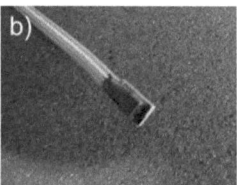 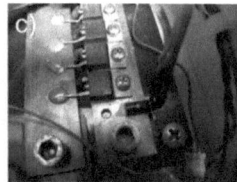

Figure 11.9.: Inset a) shows the bi-metal sensor releasing at ≈ 75° which cuts the main power line. Inset b) shows the micro-controller base temperature sensors which can be adjusted in a wide temperature range, and inset c) shows temperature sensors mounted at the high current Cu-rails in the switch-boxes.

11.4. Lower Imaging system

To optimize the MOT, respectively the optical molasses and the loading of the magnetic trap, an absorption imaging system is implemented on one axis of the MOT chamber. The imaging beam and the CCD are displayed in Fig.(11.8), were a firewire-camera DMK21BF04, with 640 × 480 pixels, and a pixel size of 5.6μm is used. The attached MD Minolta objective leads to an imaged pixel size of 22.4μm, which provides a four-fold magnification. This camera is triggered via Ethernet from a PC, and allows to take absorption images where both, the temperature and number of atoms can be measured. For further details on the imaging refer to the Appendix (A.8).

11.5. Experimental setup

The heart of the magnetic transport are without question the bunch of magnetic coils. As the currents reach values up to 150A, a water cooling is essential, as the coils must withstand the currents even for continuous operation. How the water cooling is implemented in the coils, can be seen in Fig.(11.10). The upper right insets show a vertical coil from both site of views with a copper tube directly glued onto the windings[5]. As the horizontal coils are all mounted on an upper and lower rail, the cooling is implemented on the rail rather than attached to each coil, depicted in the lowest inset. To again justify the eort of temperature sensors and water cooling, the lower right inset gives an example what better should not happen.
Inset a) gives a complete overview of the stand-alone magnetic transport scheme, showing the big ion pump in the back, four vertical transport coils and the bunch of electrical connections, next to the water cooling. Also very prominent are the firewire-camera with the attached objective in the lower left corner, and the VAT valve in the magnetic transport section as well as the two pairs of flat wires acting as horizontal compensation coils. The big horizontal coil around the center of the MOT-chamber is the upper vertical compensation coil, covering the MOT coil which can not be seen on the picture.

Even for the room-temperature transport six pairs of quadrupole coils and even six sin-

---

[5]Thermally Conductive Epoxy Encapsulant, Stycast: Emerson and Cuming, Billerica, MA 01821 - USA

11.5. Experimental setup

gle coils must be driven with the transport currents. The need for the 12 power supplies is circumvented by introducing switch-boxes which directs the current through dierent coils. The switch-boxes mainly consists of H-bridges and MOSFET benches either to simply switch on/o a coil, or to drive the current in both directions through the coil as needed for the vertical transport. Find details in Appendix (B).

switch-box	power supply	I_{max}	driven coils
1	Delta 1: 15-200	200A	push-coil, V_1
2	Delta 2: 30-100	100A	MOT-coil, H_4, V_4
3	Delta 3: 30-100	100A	H_1, V_2
4	Delta 4: 15-200	200A	H_2, V_3
5	Delta 5: 15-100	100A	H_3, V_5

Table 11.1.: A liation of switch-boxes, coils and power supplies with the capability of the dierent power supplies.

Tab.(11.1) shows the corresponding power supplies[6] and the switch-boxes, belonging to the dierent coils. Integrated demultiplexing circuits allow to control the desired coil, applying digital triggers for each coil. For further detail on the switch-boxes, containing the MOSFET-benches and the de-multiplexing circuits see Appendix (B).

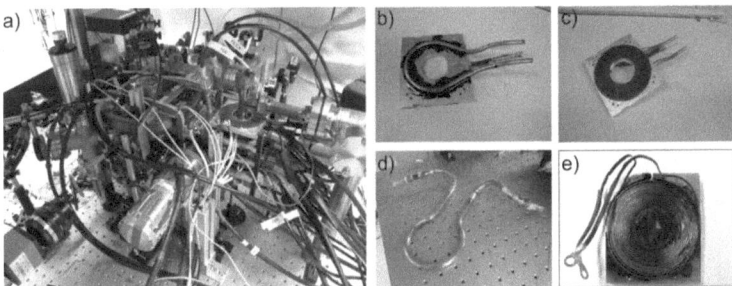

Figure 11.10.: Inset a) gives a full overview of the assembled experimental setup for the lower magnetic transport chamber. Inset b) and c) show a single horizontal transport coil H_5 in the mounting with the water cooling directly attached. Inset d) shows the main water cooling attached to the MOT-coils and the horizontal transport coils $H_1 - H_4$. A blown up transport coil of H_5 is shown in inset e).

---

[6]Delta Electronika BV, Zierikzee - NL

11. Setup of a magnetic transport line at room temperature

11.6. Experimental cycle and transport schemes

Atoms are trapped in a MOT for a loading time of $t = 10s$, at a magnetic field gradient of $18G/cm$. The cooler laser is directed onto the center of the MOT using six beams with total power of 130mW, and is operated at a detuning of $= -26MHz$ red detuned from the cooling transition $5^2S_{1/2}|F=2\rangle \to 5^2P_{3/2}|F'=3\rangle$. The repumper laser is on during MOT, molasses and optical pumping to close the cooling transition, keeping the atoms cycling on the cooling transition. After the MOT phase, the cooler is shifted further to the red, to -40MHz detuning while the magnetic field is switched o. After the molasses phase, the atoms are optically pumped to the $|F=2, m_F=2\rangle$ low field seeking state at an optimum red detuning of -50MHz, while applying a small field to define a quantization axis[7]. After the short pumping time, the repumper is switched o, to magnetically trap the atoms within 0.5ms in an intermediate quadrupole trap named magnetic trap 1 (Tab. 11.2). The trap is further compressed within 5.5ms up to a gradient of $130G/cm$. As the position of the magnetic trap and the optical molasses is perfectly matched, the trap can be ramped up rather fast.

phase	cooler laser	repumper laser	magnetic field	duration
MOT	-26MHz	ON	18G/cm	10s
molasses	-40MHz	ON	OFF	10ms
optical pumping	-50	ON	1G	1ms
magnetic trap 1	OFF	OFF	22G/cm	0.5ms
magnetic trap 2	OFF	OFF	50G/cm	5.5ms
hor. transport	OFF	OFF	130G/cm	550ms

Table 11.2.: Experimental cycle for preparing the atoms for transport

After a relaxation time in the magnetic trap of about 150ms, the transport cycle is started. The allow for dierent operational mode, three dierent transport schemes can be chosen. Atoms can be imaged in the molasses phase, or in the magnetic trap before transport. A second scheme allows to transport the atoms forth and back from an arbitrary horizontal or vertical position on the transport, imaging them in the MOT chamber using the lower imaging system (see section 11.4). This sequence for example is well suited to characterize the transport. The third protocol allows to transport the atoms inside the cryostat, and imaging them with the upper camera. Transport parameter as velocities, accelerations and the total transport time are under discussion in section (14.2).

---

[7]The quantization field is maintained via 3 compensation coils. The coils are either used for compensation of stray fields during the molasses phase, and are used to define a polarization axis applying a field in the order of several Gauss

12 Design of a 4K cryo-system

The novel experimental setup described in this thesis, includes superconducting coils and atom-traps in a 4K cryostat and a transport system for cold atoms inside this cryogenic environment. Usual commercial cryogenic systems with less demand for ultra-high vacuum, as they are closed and cryo-cooled vacuum systems, are used in solid state physics, and low temperature physics. In contrast cold atom experiments demand ultra-high and clean vacuum systems, even in a cryogenic environment, as the system is not closed and rather connected to a second vacuum chamber at room-temperature.

As even the demands for the experimental setup in regard to a realization of a hybrid quantum system including superconducting and atomic qubits are challenging, such an setup requires certain adaption of commercial cryo-systems or even a complete redesign.
The following chapter describes the design of a cryogenic cold atom experiment, were knowhow was both, painfully acquired in a hands-on approach, and on the other hand drawn from [182, 183, 184], exhibiting the extensive experiences collected in the last decades.

12.1. Cryogenic cooling systems

Related to the greek and latin originating words o and *generare*, systems are named *cryogenic* if they generate or work at low temperatures. Those systems are used in a wide range of applications, covering the whole spectrum of research, development and fundamental science.

Since the first evidence of the element helium in 1868 and the first achieved liquidification in 1908 by *Heike Kammerling Onnes*, liquid helium (LHe) with its inert chemical reactivity is still essential for cryogenic applications. As the boiling point of helium is at 4.22K at atmospheric pressure ($1 atm = 1.01325 bar$), liquid nitrogen (LN_2) is the second important element used in cryogenic applications down to 77K, the normal boiling point of N_2.

With those two liquids, cryogenic systems can be built, which cool down matter, in the easiest way by maintaining a thermal equilibrium between the test-sample and a bath of LHe or LN_2. With the development of heat exchanging machines, chiller where developed which use dierent refrigeration cycles to cool down to low temperatures.

In the following an overview of dierent cryogenic cooling systems is given, focusing on the

12. Design of a 4K cryo-system

Giord McMahon cooler used in the experiment. The operating principle then defines the properties and the ability of the machine and leads to the thermal budget which intrinsically defines design considerations of the experiment.

12.1.1. Bath cryostats

The simplest way to achieve temperatures in the range of 77K and 4K, connecting the test-sample thermally to a bath of LHe and LN2. With the advantage of cooling powers of several Watts, the biggest drawback comes along[1]: To maintain constant low temperatures, the coolant has to be refilled in time intervals which are defined by the heat load of the system and makes cold atom experiments which usually operates over dozens of hours extensive in maintenance.

12.1.2. Closed Cycle Refridgerants

In dierence to continuous gas flow cryostats using the Joule-Thomson eect or even a bath cryo-system, several systems exist where a regenerative heat exchange is maintained by an oscillating gas flow. The most prominent types are the *Stirling cooler*, the *Pulse-tube cooler* and the *Giord-McMahon cryostat* . In such cryo-systems the working gas is oscillating in a closed cycle which avoids the disadvantage of refilling the coolant. *Giord-McMahon* and Pulse-tube machines provide temperatures even down to 4K.

Pulse Tube-cooler

The Pulse-tube cryostat diers from all other closed cycle cryostat by the missing moving parts. It is a fully vibration free system, operated with a closed cycle of cold helium. It is the standard platform to go for lower temperatures in the range of 100mK and below, by implementing dilution units. In principle such a pulse-tube system would be perfect for a cold atom experiment, nevertheless its relative low cooling power and the considerable high price of the system, forces one to have a closer look on the *Giord-McMahon* cryo-cooler.

Giord-McMahon cooler

Whereas in the pulse-tube cooler a *gas-piston* is moving, the working principle of the *Giord-McMahon-cooler* rely on a physical piston, moving up and down. In analogy to electric circuits, these type of cryo-coolers can be described as RC phase shifters, where the cooling power and the heat extraction per cycle depends on the oscillation frequency of the flowing gas and the angle between the pressure and the volume within an oscillation period [183]. The dierence to a Stirling-cooler is that the gas flow is time-triggered by an inlet and outlet valve, thus maintaining a phase shift in the cooling cycle.
The Giord-McMahon cooler can in addition easily be used as a platform to reach mK temperatures driving a dilution unit attached [185, 186].

---
[1]Notice: All three cold-atom experiments done at the *Haroche-group/LKB-Paris*, *Fortagh-group/University of Tübingen* and *Shimizu-group/NTT-Tokyo* are established in cryogenic environments using bath cryostats

12.1. Cryogenic cooling systems

12.1.3. ARS closed cycle cryo-head

In my setup, a *Giord-McMahon* cryo-system GMX20-B from ARS^2 generates low temperatures at the experimental stage in the range of 3.9-4.5K. An oscillating gas flow is maintained by a moving piston, which therefore induces vibrations in the range of several millimeter of the complete system.
It provides all the advantages of relative high cooling power[3], and the closed cycle operation which reduces operation procedures to switching it on or o. The limiting disadvantage, namely the vibrations in the millimeter range, is overcome by the use of mechanical decoupling between the moving upper parts and the 4K cooling-finger. An additional cold finger covers the inner 4K cold-finger, and helium gas in between enclosed by a rubber belly, mechanically decouples [187] the vibrations similar as this is applied in pulse tube coolers. See Appendix (A.1) for more details on the performance test and the setup of the ARS closed-cycle cryo-head.

12.1.4. Thermal heat budget

In equilibrium operation, the heat load is not influenced by the helium bubble, connecting the two 4K stages, although a variation in distance between them and alternating pressure of the helium bubble instantaneously changes the temperature (see Appendix A.1).
As shown in Fig.(12.1), the temperature of the outer He-stage varies with heat load, allowing for heat input up to 1-2W if a temperature raise up to 6K at the 2nd stage is considered. Nevertheless the overall heat input should be as low as possible, facing the challenge of implementation of high-Q resonators (see section 17).
The ARS 4K closed-cycle fridge in its standard configuration just carries a tiny intermediate shield, with almost no space for cryogenic cold atom experiments. This implies that in the following several design goals are derived which the experiment has to fulfill. Based on the temperature dependent heat load Fig.(12.1), this will lead to several considerations and demands for designing and building the 4K experimental stage, the thermal shielding and the cryostat vacuum chamber, subsequently described.
The delivered and proposed standard cooling power for the cold finger DE210S implemented in the GMX20-B is therefore a first orientation which maximum heat load at the 2nd stage is allowed to maintain a low temperature in the order of 4K.

Considerations to keep the thermal heat budget low:
Regarding to the *Stefan-Boltzmann radiation* law ($P = A \; T^4$), a 300K surface radiates $\approx 50 mW/cm^2$ into the half-space, while a 50K (respectively a 35K surface) radiates $\approx 1.4 mW/cm^2$ (respectively $\approx 0.7 mW/cm^2$).

1. Assuming the cryostat vacuum chamber to maintain $A_{vac} \approx 3500 cm^2$, the surface of the intermediate shield demands to have a high emissivity in order to reduce radiation input from 300K

2. Assuming the intermediate thermal shield to maintain $A_{shield} \approx 1800 cm^2$, even the inner surface of the thermal shield and the 4K parts demands a low emissivity to prevent

---
[2] Advanced Research Systems, Inc., Macungie, PA 18062 - USA
[3] cooling power at 4.2K $\approx$ 0.5W , at 5-6K $\approx$ 2W

12. Design of a 4K cryo-system

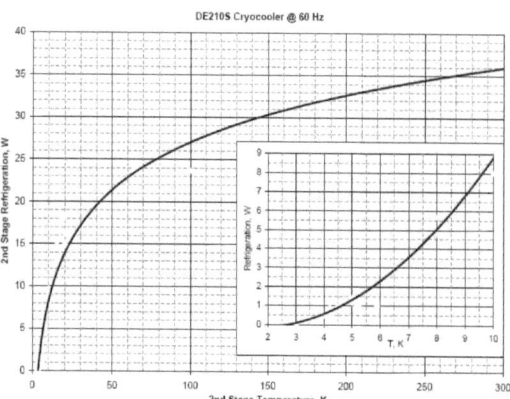

Figure 12.1: Proposed cooling power for the DE210S cryo-head implemented in the GMX20-B cryo-system. The graph depicts a rough orientation about the maximum cooling power which is allowed to maintain temperatures well below the critical temperature of Nb ($T_c = 9.3K$) [105]. The graph is taken from the performance-specification manual of the GMX20-B.

radiation towards the 2nd cooling stage and to limit the heat input drastically below 2W.

3. The current carrying wires should not provide much heating as the normal conducting wires are directly anchored to the 1st and the 2nd stage

4. The windows should be well anchored as the thermal conductivity of glasses is orders of magnitudes below those of metals

5. The vertical transport opening at the chamber bottom should just cause little heat input as the 4K parts directly *see* room-temperature

These general considerations are now discussed in more detail, as they influences the complete experimental setup.

12.2. Design goals for a cryogenic cold atom experiment

The experiment will have to provide an ensemble of cooled, trapped ^{87}Rb atoms in a cryogenic environment. It must have the capability of implementing an atom chip at $T \approx 4K$, and further must allow for microwave lines and electronics such as attenuators and amplifiers down to the 4K-stage. In addition it must provide optical access to image the atoms, and should be easy[4] to handle.

Defined by the two-inch optical windows in the thermal shield and CF63-based windows in the cryostat-vacuum chamber, a minimum distance of the experimental stage is predetermined. Retaining the coil distance $d = 30mm$ of the room-temperature vertical transport,

---
[4] As the experiment relies on a cryogenic environment, it is rather simple to achieve UHV conditions as demanded for operation with an atom-chip and cold atoms. A smart design would therefore allow frequent openings of the system to modify the setup, which would add remarkable flexibility to the experiment.

12.2. Design goals for a cryogenic cold atom experiment

this at least demands four superconducting coils.

Enough space at the 4K environment must be provided to attach an atom chip which should be easy to replace. Therefore the radiation shield between 300K and 4K must be modified at least to increase the experimental space in the cryostat, and allow to implement windows for the optical access.
In addition, the experimental sequence defines which kind of magnetic traps are needed and how the reloading onto an atom chip is achieved. Having the atoms at the final transport position, the experiment should be capable of trapping the atoms on an atom chip to allow in principle for Bose-Einstein condensation [188], and further for bringing them close to a superconducting surface, possibly close to a coplanar microwave resonator (see section 17). Following the enumeration, the experimental steps are listed:

1. Vertical magnetic transport up to a final position in the cryostat
2. Establishing a QUIC-trap with non-zero trap bottom
3. Pre-cooling for reloading
4. Transferring atoms to a superconducting atom chip
5. Bose-Einstein condensation in a superconducting micro-trap
6. Further manipulation

Alternatively, an appropriate QUIC-trap would also allow for Bose-Einstein condensation even before reloading onto an atom chip [189]. This would in addition allow to bring the atoms close to a superconducting surface/resonator-structure even without to be dependent on atom chip based micro-traps, and would increase the flexibility of the setup.

12.2.1. Superconducting vertical magnetic transport line at 4K

Regarding to section (10.3.1), the distance between the four vertical cryogenic transport coils ($V_6 - V_9$) should be similar to the vertical coils at room-temperature. The heat load in the cryogenic environment is strongly limited, nevertheless allowing to build in superconducting (sc) coils. As therefore the current is limited in the order of 1A, this will result in thousands of windings to keep the ampere windings constant, switching from room-temperature to sc-coils. Nevertheless also the dimension as the outer and the inner radius must fit with the room-temperature coils to assure similar trapping gradients, while the vertical trap gradient must also be kept constant at 130G/cm.

12.2.2. Design of a superconducting quadrupole-Ioe trap

The QUIC-trap (see section 5.3) provides a non-zero trap bottom in contrast to the vertical transport coils, which provide a quadrupole trap at the final transport position. An asymmetric Ioe-configuration trap can easily be achieved by implementation of a third coil [5] in

---

[5]The single *Ioe-coil*, which turns the quadrupole trap into a quadrupole-Ioe configuration (QUIC) trap will be called QUIC-coil.

12. Design of a 4K cryo-system

radial direction to the upper quadrupole coils. Therefore this trap allows for lower temperatures applying evaporative cooling as in a zero-bottom trap these atoms would be lost due to Majorana-spin-flips (see section 5.4). Applied currents must at least allow for a tight confinement which trap frequencies in the order of $2 \times 20Hz$ in the longitudinal Ioe-direction, and at least $\approx 2 \times 200Hz$ in both other directions to satisfy the reloading conditions shown in section (15.1).
Hence, evaporative cooling can be applied to cool the atoms perceptibly before reloading. Achieving Bose-Einstein condensation in the trap seems to be desirable, but is not necessarily demanded for reloading, and even not clear if trap frequencies of $2 \times 20Hz, 2 \times 200Hz, 2 \times 200Hz$ in the macroscopic QUIC-trap provide a trap which is capable for reaching BEC[6]. Nevertheless it would bring higher flexibility to the experiment.

QUIC-traps are known to macroscopically provide strong linear gradients, but as the trap bottom results from the dierence of two magnetic fields along the Ioe-direction, generated by two huge currents, the influence of noisy currents should be considered.
Tab. (12.1) therefore shows, how the trap bottom dramatically changes, influenced by the Ioe-current. The calculation was done for a setup with 1800 windings in the QUIC-coil, with the nearest distance to the transport axis of $d = 19.07mm$, and an inner coil radius of $R_i = 5mm$. The transport coils consist of 3000 windings with $R_i = 24mm$ and a mean distance of $D = 30mm$.

I_{quad}	I_{Ioffe}	position Δx	trap bottom	$\perp/2$
[mA]	[mA]	[mm]	[mG]	[Hz]
700	828	6.785	571	455
700	830	6.770	798	391
700	832	6.775	1025	316
700	834	6.742	1252	311

Table 12.1.: Influence of the Ioe-current in a QUIC-trap on the trap bottom and the transversal trap frequencies

Therefore it is necessary to apply the same current in series on all three coils, as at least the position of the trap stays constant if $I_{Ioffe} = const. \times I_{quad}$. Nevertheless this would constrict the degrees of freedom in the trap, as than the trap bottom is directly coupled to the position, whereas the trap frequencies can be increased, of course on the cost of the trap bottom.
A possibility to partially overcome this restriction is to introduce a *super-Ioe-coil* as shown in Fig.(12.2). This coil will be driven independently from the serial circuit of the QUIC-trap which performs I_{quad} and $I_{Ioffe} = const. \times I_{quad}$, and allows at a certain position and trap bottom, to independently fine-tune the parameter, with a much smaller influence, as the tuned super-Ioe current is quite small.
It is an intrinsic property, that the atoms are shifted by a certain position away from the transport axis, towards the Ioe-coils This oset position must be considered implementing a micro-trap on an atom chip surface, as this is the starting position for the reloading-process (see section 15.1) onto the atom chip.

---
[6]The important parameter would also cover, lifetime in the trap and phase space density

12.2. Design goals for a cryogenic cold atom experiment

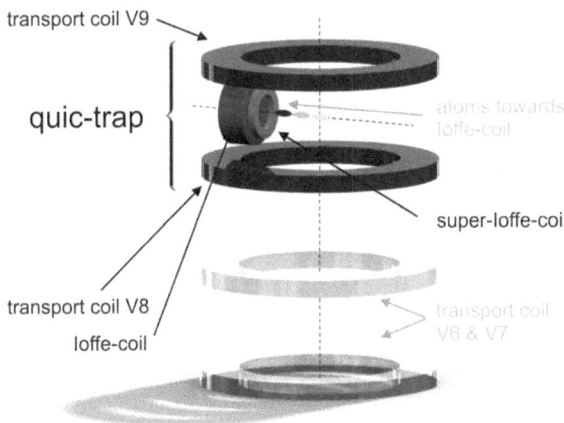

Figure 12.2: This design-drawing shows the way how a quadrupole-Ioffe configuration (QUIC) trap can be implemented. The additional super-Ioffe coil allows for a further degree of freedom while increasing trap stability. Establishing the QUIC-trap, the atoms are pulled towards the Ioffe-coils away from the axis by Δx.

12.2.3. Superconducting micro-trap

A superconducting (sc) micro-trap even allows to provide stronger confinement than in a macroscopic Ioffe-configuration trap (see section 5.3.4). With the demand of bringing the atoms close to a superconducting surface/MW-resonator, a micro-trap implemented on an atom chip, would therefore be a well suited tool. Several studies [103, 104, 105], highlight the different properties of sc-wire traps as in superconducting structures the *Meissner-effect* [190] plays a role, changing the current distribution in the wire. Nevertheless this effect plays only a role if the atoms come really close to the trapping wire, in the order of its dimensions. Regarding the reloading of atoms from a QUIC-trap onto an atom chip with its implemented Z-shaped Ioffe-trap, it is sufficient to contemplate the far-field of the sc-chip-trap, which still can be calculated *classical*.

Even to maintain a chip based micro-trap, additional homogeneous fields must be applied which build up a Ioffe-like trap with the Z-current (see section 5.3.4). This is done by the so called bias-coils and the bias-Ioffe-coils, both in the chip-plane. An additional vertical pair of coils would allow to tilt the trap, applying a homogeneous vertical field.

12.2.4. Atom chip mounting

The atom chip provides the experimental platform to combine atomic- and solid state physics. It is well suited to implement microscopic wire traps to manipulate atoms, and for superconducting coplanar MW-resonators. As the chip is in the 4K environment, and high quality coplanar MW-resonators/wave guides demand superconducting structures to reduce losses

12. Design of a 4K cryo-system

and increase fidelity, it is made of niobium[7]. Hence the chip must be well anchored to cool it down, and lead o radiation input. The chip mounting must therefore provide a good thermal contact between the sc-chip and the 4K environment, where eddy-currents should be avoided by all means. Together with a flexible design, which allows to replace it without di culties, a non-metallic mounting is favored.

12.2.5. Current wires down to 4K

The main heat source beside the thermal radiation (see section 12.1.4) would be the heat dissipation through the wires, which connects all the magnetic coils to the outside-world. It is obvious that the heat load down to the 2nd cooling stage depends just on two, countervailing e ects, namely the thermal conductance through the wire cross section of the Cu-wire, and the heat dissipation in the normal conductive part if a current is flowing. While the first depends on the cross section and the length as

$$P_{cond} \propto \;(T) \times \frac{A_{wire}}{l_{wire}} \quad (12.1)$$

the dissipation scales with the ohmic resistance as

$$P_{diss} \propto \;_0(T) \times \frac{l_{wire}}{A_{wire}} \quad (12.2)$$

leading to a total heat load weighted with $\;_{cond}$ and $\;_{diss}$[8] which reads as

$$P_{total} = c1 \;_{cond} \;(T) \frac{A_{wire}}{l_{wire}} + c2 \;_{diss} \;_0(T) \frac{l_{wire}}{A_{wire}} \quad (12.3)$$

where in Eq.(12.1) and Eq.(12.1), temperature dependent material parameter enters. Therefore optimum wire parameter (l_{wire} and A_{wire}) can be found for the di erent current carrying wires if a typical current load is assumed[9]. In addition the length of anchoring wires also influences the heat load at the cold stage [191]. In the following di erent wires are listed which will be demanded to connect the above mentioned coils.

For the di erent purposes, the desired coil-current depends on the functionality of the coils. Tab.(12.2) therefore sums up the total heat load partially caused by the Joule heating and the heat conduction through AWG28[10] wires for a e ective length of $\approx 80cm$ from 50K to 4K[11] summing up $P_{cond} + P_{diss}$. From the desired magnetic field or field gradient for a certain geometry, the desired current is roughly estimated following [192] within a certain degree of freedom, to ensure full experimental flexibility, assuming for the coils to participate continuously as it would happen in a worst case scenario. This finally leads to the estimation for the demanded anchoring length, or optimal length of the current carrying wires. The temperature

---
[7]Niobium is well suited as it can be micro-fabricated and provides the highest critical temperature in single element superconductors
[8]The heat power due to conduction is a permanent load, while for dissipation in an approximation the time of the duty cycle of the coil comes into play. For a worst case scenario, the coils are assumed to be switched on continuous.
[9]Current load in this manner means both: the maximum current strength and the fraction of experimental cycle in which the current participates in heat input.
[10]American Wire Gauge 28: $d = 321 \mu m$
[11]The worst case is calculated with RRR=100 at $\chi = 10^{+7}$ and a corrected wire temperature from 50K to 4K

12.2. Design goals for a cryogenic cold atom experiment

Functionality of coil	B-field parameter	Nominal current I_N	# of leads	$\mathcal{L}_{diss}$	$\mathcal{L}_{cond}$ (50K→4K)	Total heat load [mW]
Transport coil	130G/cm	1A	8	1	1	8× 19.8mW
QUIC-coil	130G/cm	1.5A	2	1	1	2× 32.0mW
Super-Ioe-coil	$> 5\% B_{QUIC}$	300mA	2	1	1	2× 10.8mW
RF-coil	—	200mA	2	1	1	2× 10.3mW
Bias vertical	100G	3A	4	1	1	4× 98.5mW
Bias Ioe	30G	1A	2	1	1	2× 19.8mW
Bias chip	60G	1A	2	1	1	2× 19.8mW
Chip wire	—	2A	2	1	1	2× 49.9mW
Total heat load						**≈ 840mW**

Table 12.2.: Overview of the heat budget for the wiring

sensor connectors therefore are completely neglected in the heat budget.

With the heat input depending on varying material properties $P_{cond} = \kappa(T) \Delta T \mathcal{L}^{-1}$ with $\mathcal{L} = l_{wire}/A_{wire}$, a total thermal conductivity integral can be introduced which than leads to

$$P_{cond} = \frac{\mathcal{L}_1 - \mathcal{L}_2}{\mathcal{L}} \quad (12.4)$$

already accounting influence for the temperature onto the thermal heat conductivity and with the dissipative heat input

$$P_{diss} = \rho(T) I_N^2 \mathcal{L} \quad (12.5)$$

which accounts for a temperature dependent ohmic resistance $\rho = \rho_0(1 + \alpha \Delta T)$. For simplifications the temperature is assumed to decrease linear with distance in the wire.
For the optimization of the wire length and the wire cross section, the best parameter are derived from the above given estimation Eq.(12.4) and Eq.(12.5), which shows how the minimum can be found. As the left branch of the curve Fig.(12.3) is due to the heat conduction through the wire, the right branch origins from the ohmic dissipation as the ratio $\mathcal{L}$ increases.

12.2.6. Good optical access

Optical access is unavoidable if the atoms should be optically imaged, as long as the imaging laser and the CCD are not implemented inside the cryostat. This line of sight would introduce the need of windows at the radiation shield, and implies that attention has to be paid to the anchoring of the windows, which must as good as possible be anchored to the thermal shielding as the introduced heat radiation goes as $P_{window} \propto A \times T^4$ according to Stefan-Boltzmann. In addition a proper imaging in further experiments demands a well defined polarization of the incoming laser beam. Birefringent windows would therefore turn the polarization which than leads obviously to an unknown polarization, leading to an underestimation of the number of atoms as described in Appendix (A.8). Nevertheless accurate imaging and estimation of atom number will some when be crucial, and therefore it is not allowed that windows induce

12. Design of a 4K cryo-system

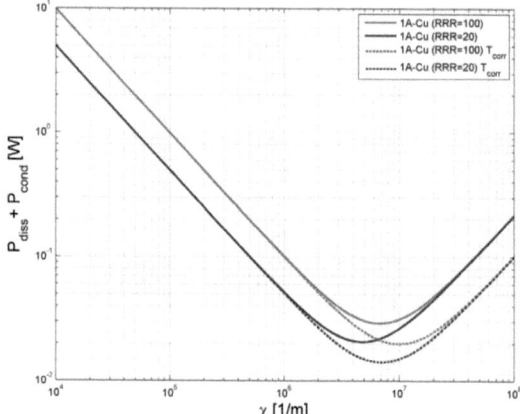

Figure 12.3.: For two dierent Cu-wire types with RRR=100 and RRR=20, the total heat load down to the 2nd cooling stage at $\approx 4K$ is estimated. Even if the temperature devolution in the wire is not known exactly, the dashed curve assumes the specific resistance to depend on a linear decreasing temperature with wire length, whereas the solid line assumes the ohmic resistance to be constant at $(T) = const. = (50)$. Estimation is done for an AWG28 wire with 1A continuous flowing. The reduced, inverse length is defined $= l_{wire}/A_{wire}$

a phase shift between the orthogonal polarization directions. Following [18], it should be fine if the birefringence of the windows is small enough that the polarization shift through the whole system is below or in the order of 0.3 rad, which seems to be suitable and practicable.

12.2.7. Fast switching of SC coils

The demand for fast switching of the superconducting coils is threefold. It basically relies on the following three questions. A) How fast can the coils follow the applied current, as this would limit the vertical transport speed? B) How fast can the currents be changed, regarding to reloading of the atoms from one to another trap, or just changing the trap parameter? C) How fast can all magnetic fields be set to zero, to let the atoms expand during free fall to apply time-of-flight (TOF)[12] imaging?

By intrinsic properties, superconducting coils have almost zero resistance, as just the feed lines at higher temperatures contribute to the coil-resistance. Due to their high number of windings, the coils have a huge inductance compared to the room-temperature transport coils following

---
[12]For further detail on the method of detection see Appendix (A.8).

$$L = N^2 \ R_m \ \mu_0 \left[ln\left(\frac{8R_m}{a}\right) - 2 \right] \quad (12.6)$$

where $R_m = (R_o + R_i)/2$ is the mean radius of the coil, N the number of windings and $a = R_o - R_i$ the width of the coil regarding to [192]. For the described setup of the superconducting transport coils (see section 13.4) this can even result in an inductance of several hundreds of mH.

Switching o the inductive coils results in an exponential decay, were the time constant is defined by $= L/R_{4K} \approx 1s$, with simultaneously generating a dramatic induction voltage, depending on the switching time[13]. Both can be overcome by introducing a quenching circuit (see Appendix B), which linearly sets down the current in $t_{quench} \approx ms$, and limiting the induction voltage.

Nevertheless, eddy-currents can occur following the induced electric field E and hence the induced voltage in a closed metallic loop described by

$$\oint E \ dl = -\frac{d}{dt} \int_{surface} B \ ds \quad (12.7)$$

caused by a switching magnetic field $\partial B/\partial t$. Such closed loops were the eddy-currents can flow are provided by parts of the experimental stage and Cu-mountings, induced by the switch-o of near by coils. It is therefore of outstanding importance, that such closed loops in metal parts as good as possible avoided[14]. In addition eddy-currents cause ohmic dissipation and therefore additional heat in the metal loops, as well as they induce persistent currents in closed superconducting loops. Hence, the use of dierent metals has to be weight, as they should provide good thermal conductivity while having a high electric resistance to suppress the amplitude of eddy-currents.

12.3. 4K experimental stage

12.3.1. Design considerations for the 4K stage

The 4K closed cycle cryo-finger, provides just a rather small 4K surface. Therefore a 4K experimental stage must be implemented, capable of gathering a bunch of mountings. It is the platform, where the wires are anchored before connecting them to the sc-coils and were temperature sensors are connected. In addition it maintains the coil mounting for the sc-transport coils, and the chip-mounting. To account for a good thermal conductivity it should be build of high purity copper, possibly goldened to increase reflectivity, and prevent oxidation which would reduce thermal contact.

---

[13] With $L \approx 400 - 700mH$ and $R_{4K} \approx 0.5$ caused by the high temperature connection wires, the free-run switch o time would be in the order of seconds.

[14] Eddy-currents were reported to cause major problems in superconducting cold atom experiments as they were done at NTT in Japan [19] and at ENS/LKB in Paris [18].

12.3.2. Connectors

To allow for high flexibility the connections between superconducting coils, chip-wires and the temperature sensors, should be easily removable. As the contacts between the superconducting and normal-conducting wires introduce a critical interface, were heat can be dissipated into the sc-wires e.g. due to solder-contacts, they demand extraordinary attention even if they should be designed as connectors. A detailed review on this considerations is shown in Appendix (A.3.2). It is obvious that a connection between a sc-wire and a normal-conducting wire with two much heat dissipation can easily lead to a breakdown of superconductivity as an avalanche starts if the dissipated heat is not lead away fast enough.

12.3.3. Coil-mountings

Above considerations lead to a rather sophisticated design of the coil-mountings sustaining the superconducting trapping coils. They are made of high purity, oxygen free high conductive, goldened copper (OFHC-Cu)[15], to assure good thermal contact and high thermal conductivity. In parallel they form a closed electric loop which would generate eddy-currents, switching of the coils. To circumvent this drawback, the coil mountings should be sliced adequate to reduce the closed loop area.

12.4. Thermal shielding

The heat load performance of the cryo-system shown in (Fig.A.2) implies, that the room-temperature radiation from a 300K surface should be shielded from the 4K surface. In this context the thermal shielding has to account for two things. A) good thermal conductivity, but a not too small intrinsic electric resistance to prevent eddy-currents and B) it must be capable of the window-mountings which allows for optical access.
To ensure the demanded functionality of the shield the design considerations are discussed in the following.

12.4.1. Anchoring of windows at low temperatures

As already mentioned in section (12.2.6) the mounting of the window is crucial, both for the non-birefringence and for a good thermal contact as the radiated heat load should be as low as from the thermal shielding at $\approx 50K$ and not higher because of a bad thermal conductivity of the windows. To reduce tightening stress, dierent mountings could be applied, using indium seals [193, 194] or even dierent adhesives [195, 196, 197] or Teflon based mountings [198] to fix the windows, regarding to well established techniques. An almost strain-free method [199] could not be implemented, as it does not provide stability and best thermal contact. A detailed realization of the window mounting, dierent trials and errors and the accompanying issues can be found in Appendix (A.2).

12.4.2. Design considerations for a cryo shield

As the radiation input power goes with the surface of the shield, a bigger shield would also lead to a higher radiation input at the 1st cooling stage, under same conditions. To even

---
[15]For material properties see Appendix D

increase the performance of the shield it is worth to think of a thermal insulation instead of a reflection coating, accounting the emissivity rather than the reflection coefficient. As the emissivity is defined as the ratio of the energy emitted by a particular material, divided by the energy emitted by a black body at the same temperature, the absorbed heat load at the thermal shield from the vacuum-chamber can be estimated.

$$\epsilon_{1,2} = \frac{1}{\frac{1}{\epsilon_1} + \frac{1}{\epsilon_2} - 1} \qquad (12.8)$$

The efficiency which describes the absorption of radiated power at the thermal shield from the outer vacuum chamber therefore is lowest if both, $\epsilon_1 = \epsilon_{chamber}$ and $\epsilon_2 = \epsilon_{shield}$ is a minimum.
A possible way to realize such a thermal insulation [200] would therefore consist of several layers of super-insulation foil[16] separated by a plastic net while the net also alleviates the evacuation of the inner layers, if this super-insulation is put into a vacuum-chamber.

12.4.3. Material and heat conduction

Different materials can be used for a radiation shielding whereas high purity aluminum is best suited. It combines high thermal conductivity[17], even not as good as ECu or OFHC-Cu with a high RRR[18] value, but has an electric resistance which is not as low as in Cu, reducing the sensitivity for eddy-currents. Aluminum also provides a good mechanical properties which make it easier to machine, as high purity Cu is even too soft to be machined accurate.
The most relevant question is, if there is a difference in the cooling time to bring down an Al or Cu shield from room-temperature to at least several dozens of Kelvin, considering various shield thicknesses. Fig.(12.4) shows a comparison for Al3003F[19] and Cu with RRR=3[20]. This Cu acts therefore as a lower bound, as high purity Cu provides RRR up to 2000. The cooling time is estimated assuming 10% of the radiated power at 300K to be absorbed at the thermal shield, which is connected to a 55K thermal bath. A rough estimation of the time scales can than be derived, if a geometry similar to the thermal shield (see section 13.2) is assumed[21]. The important conclusion from Fig. (12.4) therefore is, that for different shield thicknesses of 1, 4, and 8mm, the timescale just depends on the used material.

12.5. Cryostat vacuum-chamber

The demands for the cryostat vacuum chamber more or less depended on the flexibility of the setup. It should provide optical access for imaging of the atoms, whereas at least the

---

[16] A super-insulation consists of a thin plastic layer which is aluminized to lower emissivity. It is commercial available at cryogenic companies and well known in the community since decades [182, 183]
[17] For material properties see Appendix D
[18] Residual Resistivity Ratio (RRR): gives the ratio of the electrical resistance between 300K and 4K. A high RRR-value is a measure for high purity and good thermal conductivity.
[19] Thermal properties for this Al alloy can be found under the UNS number A93003.
[20] $\lambda_{T=50K} = 200 \frac{W}{m\,K}$
[21] Notice: this is a rough estimation, as neither the thermal conductivity nor the cooling power of the cryostat is assumed to be temperature dependent.

12. Design of a 4K cryo-system

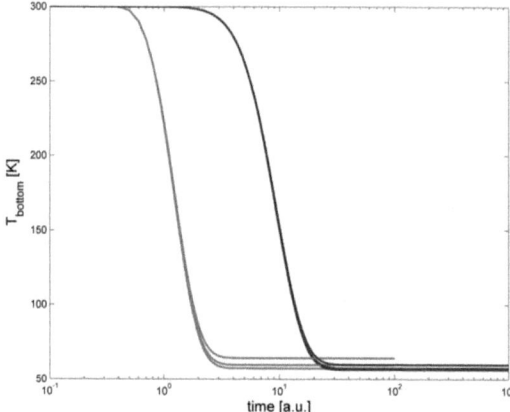

Figure 12.4.: The fast cooling exhibits the performance and the bottom-temperature after during cool-down of Cu with RRR=3, which is a rather bad Cu and gives an upper bound for Cu. The right curves (blue) are for aluminum Al3003F. Both materials are calculated for three dierent shield thicknesses: d=1,2,4mm for the thermal shield. The consequence is that the performance rather depends on the material than on the shield thickness.

diameter of the windows should not limit the numerical aperture provided by the windows in the thermal shield or the inner setup.

Further, a bunch of several additional flanges should allow to attach pumps and feed-throughs for wires, whereas a CF16 flange at the bottom acts as the connection to the vertical transport section. Considering the finite thickness of a thermal shield, the spacing between the shield and the chamber, as well as the thickness of the flange, attention must be paid to comply with the distance between the last outer and the first inner transport coil.

To avoid permanent magnetization due to the bunch of fast switching transport coils, the chamber should be made of low magnetic steel. At least the lower and the top flange must be realized rotatable to align the two chamber, the cryo-vacuum chamber and the magnetic transport in respect to each other.

13 Setup of a 4K cryo-system

13.1. The Cold finger and the Vacuum system

The cold-finger, delivered from ARS[1], came along with a rather small thermal shield as shown in Fig.(13.1). It is flanged with its CF160 onto an adapter flange, a CF200, which fits on the vacuum chamber, inset b.). The vacuum chamber itself as shown in the drawing, has at least two optical axis, on each mounted a pair of CF40 and CF63 windows, and a third axis consisting of a CF16 and a CF40 flange in opposite under an angle of 45°. The four cones through the windows depict the numerical aperture of the setup.

At the top of the vacuum chamber, 4 CF40 flanges allow to attach pumps or throughs to keep the setup flexible. Both, the lower and the upper CF200 flange are rotatable to align the transport chamber, and the cryo-finger to defined axis. Inset c) on the right shows the cold-finger attached to the vacuum chamber, with the CF160-CF200 adapter flange and the rubber bellow between the two square plates, which allows for mechanical decoupling of the system.

This mechanism, which allows for vibrational decoupling is implemented in the setup by mounting the upper cold head on the ceiling, whereas the outer cold finger with its several cooling stages is mounted on the optical table.

13.2. The radiation shield

The thermal shield for radiation protection was designed and built for the experimental demands (see section 12.4). Compared to the delivered, small shield as shown in Fig.(13.1), inset a), the shield is much bigger, capable of the 4K experimental setup and four window-fittings on the two optical axis. It consists of a 4mm thick tube of high purity aluminum (99.7%), and is a two component shield with an uppper and a lower part. Thermal contact between the upper and the lower shield aswell as to the 1st cooling-stage are improved by a silver foil[2].

---

[1] Advanced Research Systems, Inc., Macungie, PA 18062 - USA
[2] Silver-foil is used as the soft Aluminum is warped as the screws are often tightened. In addition the Ag-foil provides a better thermal contact than vacuum grease

13. Setup of a 4K cryo-system

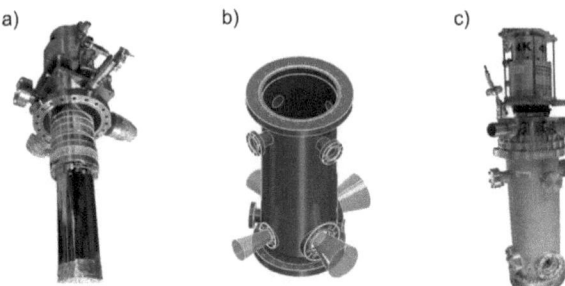

Figure 13.1.: The two major components of the cryogenic setup are shown. Inset a) shows the cold-finger with a small thermal shield attached at the bottom, and inset b) shows a drawing of the vacuum chamber made of 316LN low magnetic steel. Both components are attached via an CF160-CF200 adapter flange

The connector rings are welded onto the tube and contain 16 bores for M6 screws, to thermally connect the two shield components. Fig.(13.2) shows the different components of the shield, whereas inset a) shows the bare Al-shield after manufacturing. To lower the emissivity of the surface, the shield is insulated using a low emissivity-foil[3], and a polyester netting[4] in between to keep the different layers in distance. Inset b) shows one of the net-layers, whereas inset c) shows the already fully insulated bottom part of the shield insulated with 12 layers of the low-emissivity foil. In addition also the inner part of the shield was treated with an Al-foil with high reflectivity[5], to lower the emissivity on the inner surface of the shield.

For test-reasons the window-mountings are closed with Al-blindings and held with Teflon-stripes[6], as just in the late phase of the setup the windows were implemented. These windows are 4mm thick with 50mm diameter made of SF57[7], with an anti-reflection coating at 780nm for an incident laser-beam, under $0°$. The manufacturer therefore ensures a reflection coefficient $R < 0.25\%$.

The chosen SF57 windows were used, as they already proved of value as reported in [18], regarding the implementation of non-birefringent windows (see Appendix A.2) which are necessary to apply accurate absorption imaging (see Appendix A.8) of the cold trapped atoms.

Inset d) of Fig.(13.2) shows the complete thermal shield mounted on the cold-finger. In the back of the picture one can see the vacuum chamber as the cold-finger was removed from the chamber for re-building reasons.

---

[3]Superinsulation NRC-2, from ICE Oxford
[4]ICEles45 polyester netting from ICE Oxford
[5]ICEles43 - aluminum Foil from ICE Oxford
[6]These Teflon-stripes are in the later experiment replaced by a Teflon ring and an additional steel-ring to press the windows/blindings onto the window-mounting.
[7]SF57 anti-reflection coated windows ($R_{\lambda=780nm} = 0$) from Laseroptik GmbH

13.3. Inner life of the cryostat

Usually the thermal shield was operated at 51-53K, whereas for the commercial shield in test-runs temperatures down to 31K were achieved[8]. Measurements of the temperature at the SF57 windows showed that they even remain at higher temperatures and do quite bad thermalize with the shield as temperatures above 100K were measured on the windows. Hence, some eort was put into the right design and fixing-technique of the windows and window-mountings as some of the windows broke during test-runs and did not thermalize well with the shield. Find more details in Appendix (A.2).

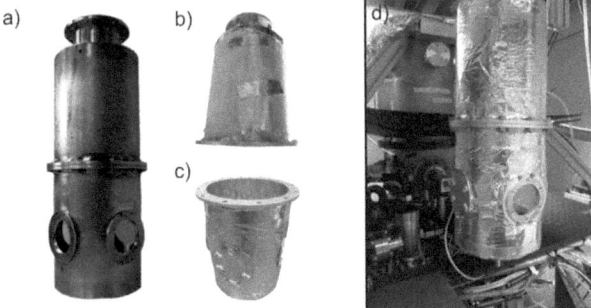

Figure 13.2.: The setup of the high purity aluminum shield, inset a) is shown, which protects the 4K stage from thermal radiation. Inset b) and c) show the top and bottom part of the shield with its thermal-insulation layers. The shield is mounted on the cold-finger and shown during rebuilding in inset d) in front of the vacuum-chamber.

Inset d) of Fig.(13.2) also exhibits the small Al-tube at the bottom of the thermal shield which acts as a dierential pumping stage between the 4K area and the transport section. It fits in the CF16 vertical transport tube and also provides three Kapton-rings which additional separates the outer cryo section between the 50K shield and the chamber-wall from the transport section.

Several 12mm diameter holes are drilled into the top cap of the thermal shield to partially connect the 4K environment and the outer cryo-section which is pumped by a turbo-molecular pump. This enables to pump the experimental part and the vertical transport section in the early phase of evacuation. Nevertheless all holes, except 2 of them are closed with Al-tape as it turned out that the holes were responsible for bad vacuum in the 4K-section (see Appendix A.5).

13.3. Inner life of the cryostat

This section gives an overview of the implementations on the 4K stage, such as the temperature sensors, the wires down to 4K and the base-plate.

---

[8] This might be due to the much bigger surface of the self-made shield and some thermal bottlenecks resulting from the weldings, compared with the small shield.

13. Setup of a 4K cryo-system

Wires

The superconducting coils and the RF-coil are connected via a set of wires reaching from 300K down to 4K. In addition further wires are installed to connect the temperature sensors and possibly connect further devices, such as the atom chip and microwave devices.
Two feed-throughs at the cryo-head maintain 20 pins for wires each. Some of them are used to connect the four temperature sensors and at least 6 pairs of wires are used to connect the four superconducting transport coils, the Ioe- and the super-Ioe-coil. The dimensions of the wires were optimized to minimize the heat input and the thermal dissipation at the 4K stage. Tab.(13.1) gives an overview of the different wires, and the further use of them.

Feed 1+2, Pin No.	d_2 Feed 1	d_2 Feed 2	Feed 1: used for	Feed 2: used for
A	0.113mm	0.321mm	sensor T_C	not used
B	0.113mm	0.321mm	sensor T_C	coil V_9
C	0.113mm	0.321mm	sensor T_C	coil V_8
D	0.321mm	0.321mm	not used	coil V_6
E	0.321mm	0.321mm	not used	RF-coil
F	0.321mm	0.321mm	not used	super-Ioe
G	0.321mm	0.321mm	not used	not used
H	0.113mm	0.321mm	sensor T_D	not used
J	0.113mm	0.321mm	sensor T_B	coil V_7
K	0.113mm	0.321mm	sensor T_C	coil V_9
L	0.113mm	0.321mm	sensor T_B	coil V_8
M	0.113mm	0.321mm	sensor T_B	super-Ioe
N	0.113mm	0.321mm	not connected	coil V_6
P	0.321mm	0.321mm	not used	RF-coil
Q	0.321mm	0.321mm	not used	not used
R	0.113mm	0.321mm	sensor T_D	not used
S	0.113mm	0.321mm	sensor T_B	coil V_7
T	0.113mm	0.321mm	sensor T_D	Ioe coil
U	0.113mm	0.321mm	sensor T_D	Ioe coil
V	0.113mm	0.321mm	not connected	not used

Table 13.1.: Wires installed from $50K \rightarrow 4K$. The dimension of the wires was optimized to minimize heat load down to 4K

At Feed-through 1 (Feed 1), the wire diameter from 300K to 50K is 0.5mmm, whereas at Feed 2 the used Cu-wire provides 0.321mm diameter. The second stage is connected via an AWG37 ($d = 0.113mm$) or an AWG28 ($d = 0.321mm$) Cu-wire[9] as shown in Tab.(13.1). Both lengths of the Cu-wires at the first and the second stage are 1.9m. Also shown are the labels of the three temperature sensors which are installed at the 4K environment, whereas the fourth sensor is connected via a third feed-through (CF16) to the thermal shield.

Fig.(13.3) shows how the wires from 300K down to 50K are wounded around the 1st cooling-

---
[9]AWG: American Wire Gauge

13.3. Inner life of the cryostat

stage and thermally anchored at the Cu-block. Care was taken about a space-saving implementation, as up to 40 wires have to be carefully wounded and anchored on the limited Cu-space, taping them with Al-tape[10] for better tightening. Inset a) shows the wires at the first cooling-stage before soldering them together with the wires down to the 4K stage. Inset b) shows the insulation tubes around the solder contacts[11] and the twisted pairs going through a slit in the Cu-block of the 1st cooling-stage, down to 4K.

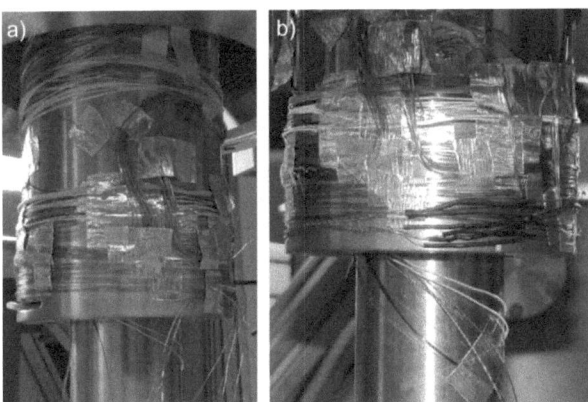

Figure 13.3.: Inset a) and b) show the normal conducting Cu-wires at the first cooling-stage of the cold finger. The wires are accurate wounded around the Cu-block and thermally anchored wire tightening the windings to the surface using Al-tape as shown in the figure. Above this, Teflon-stripes are wounded to again tighten the windings, while the surface is than protected from room-temperature radiation taping it with a high reflectivity aluminized plastic tape, not shown in the figure. The black heat shrink tubes prevent shorts between ground and the solder-contacts of the wires.

4K Base-plate

To provide better access to the 4K base-plate where most of the 4K wire contacts are thermally anchored, an adapter stage ($= 60mm$) was introduced, connecting the base-plate and the bottom of the 2nd cooling-stage. Fig.(13.4) shows this adapter stage insulated with a white Teflon stripe. During the setup of the experiment dierent types of base-plates and several kinds of connections for the wires were tested. Finally realized was a setup as shown in Fig.(13.4), exhibiting three bobbins which acts as thermal anchors for solder contacts to the superconducting transport coils. A 9pin sub-D connector maintains two of the temperature sensors. This type of connection therefore allows for uncomplicated removal of the 4K interior to allow flexible rebuilding of the experimental stage. Nevertheless it turned out, that

---
[10] Al-tape ICEles14 from ICEOxford
[11] For all solder contacts a solder with 39% lead is used: Type HF32, S-Sn60Pb39Cu1. The contact itself is protected via a heat shrink tube and is super-conductive at $T \approx 7.5K$

13. Setup of a 4K cryo-system

the connection between the normal- and the super-conducting wires must be done by solder contacts, properly anchored to the 4K plate.

The base-plate also carries three char-coal container made of aluminum and covered with an Al-grid, to keep the char-coal pellets[12] in the container. The char-coal containers are implemented to increase the pumping speed and to allow for better vacuum in the 4K-environment[13].

As the experimental setup of the 4K stage is rather challenging, connecting and anchoring almost 40 wires and implementing solder-contacts to superconducting wires, quite some issues occurred which are for completeness and in detail described in (A.3).

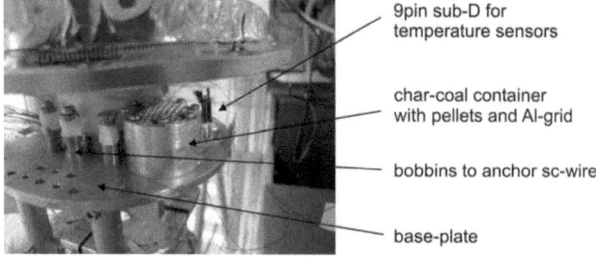

Figure 13.4.: 4K interior with 9pin Sub-D connector, char-coal container and heat-sink bobbins on the golden experimental 4K-platform. Cu-parts are goldened to increase reflectivity and to provide good thermal contact between surfaces.

Temperature sensors at 4K

To allow for advanced temperature control in later experiments a heating stripe is implemented at the bottom of the second cooling-stage. This, and nevertheless the request for full experimental control demands temperature sensors. Two carbon ceramics sensors (CCS)[14] are implemented at the 4K environment, while the third sensor at 4K is a magnetic field independent calibrated Cernox[15] sensor. They all diers in the type of sensor-mounting as shown in Fig.(13.5). Inset a) shows sensor $CS3$ which is a CCS, mounted on a Cu-bobbin at the base-plate. Inset b) is the Cernox gauge-sensor which is located at the lowest transport coils and provides a stability in the sub-mK region and inset c) shows the Tab-D sensor which is mounted on the uppest transport coil. The CCS as well as the $CS3$ slightly depends on magnetic fields[16]. All three sensors were gauged down to 4.3K and are monitored via a

---

[12] 3mm steam activated charcoal pellets from Carl Roth GmbH, Karlsruhe - D
[13] Compared to the demands for UHV, the 4K environment would never allow for low pressures at high temperatures as things are *dirty* and can not be baked out. Solely the low temperature of this experimental stage allows to go for low pressures. This should be enhanced by the char-coal, as the 4K environment is not a closed system and also *sees* regions with higher pressure
[14] Carbon-ceramic-sensor (CCS) for $T = 1.6 - 300K$, uncalibrated from SCB, Service and Consulting Cornelia Budzylek
[15] Lakeshore: type 1050 with $\pm 3mK$ accuracy at 4.2K and $\Delta T/T_0 \approx 0.1\%$ at B=2.5T
[16] They dierence in stability if the magnetic fields for transport are on/o is less than 2%

13.3. Inner life of the cryostat

temperature controller.

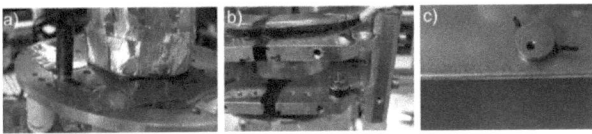

Figure 13.5.: Dierent temperature sensors are shown which are mounted in the 4K environment

Overview of the 4K environment

The following figure gives an overview of the implementations at the 4K environment. It shows the base-plate with the vertical attached goldened[17] rods which sustain the four Cu-mountings for the vertical transport coils as shown in inset a) of Fig.(13.6). Inset b) shows one sliced 4K blinding between the upper two transport-mountings which protects the coil from thermal radiation from the shield, and two char-coal container mounted bottom-up and bottom-down at the base. On the backside of the sliced Cu-plate the RF-coil is mounted which consists of a normal conducting Cu-wire, 20 windings in a diameter of 20mm, used to shine in a radio-frequency signal applying evaporative cooling of the atoms in the magnetic trap.
The two uppest coils as well as the Ioe- and the super-Ioe coils are surrounded by an Al-tape (for details see Appendix A.3).
Inset c) shows the design-drawing of the complete 4K-environment. In addition to the implementations shown in inset b), the figure exhibits bias coils for homogeneous magnetic fields as they will be used for the chip-based micro-traps.

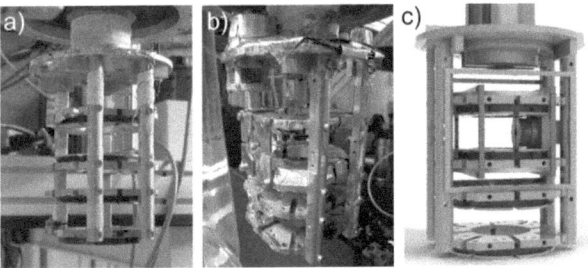

Figure 13.6.: 4K setup with just vertical transport coils implemented (inset a), and the experimental setup as used for establishing the atoms in a superconducting QUIC-trap, inset b). Inset c) on the right shows a design-drawing with three dierent rectangular/circular bias-coil pairs for homogeneous magnetic fields. They are colored red, gray and green.

---

[17] All implemented Cu-parts at the 4K stage are coated with $2\mu m$ gold using an electrolytic method.

13. Setup of a 4K cryo-system

This figure in addition highlights all dierent ways to fix things together. With priority Cu-parts are screwed together as this provides the best thermal contact. Two dierent methods rely on low temperature Varnish[18] gluing things together, whereas the third method either rely on Al-tapes or a Teflon tape to tighten things together, aiming for a good thermal contact.

13.4. The superconducting coils

It took more than five month starting with the first test coil, until all four transport coils achieved the nominal current in the superconducting state. Applying a learning-by-doing strategy the best way for building the transport coils was figured out, making them all run. A complete discussion on the sc-coils can be found in Appendix (A.3).

13.4.1. The vertical transport coils

Dierent types of coils and mountings are implemented in the setup. They dier in the mounting geometry and the way how the coils are attached to the mounting. The Cu-mountings are extensively sliced to provide just small closed metal loops, reducing eddy-currents. In addition they must provide extraordinary good thermal contact to the coil. For winding the coils, two dierent ways are applied considering the two mounting geometries either with a cap, shown in Fig.(A.5), inset A), or without a cap, inset B).

Slot	transport coil	mounting type	coil type
1	V_6	without cap	D09
2	V_7	with cap	B10
3	V_8	with cap	B03
4	V_9	without cap	C05

Table 13.2.: Dierent coil, and coil-mounting types for the sc-vertical-transport

The used wire[19] consists of a Cu-matrix with 25 niobium filaments. If below the critical temperature T_c for Nb, the filaments maintain the flowing current, while the Cu-matrix also endures the current if T_c is crossed, and ensures heat to be drained o.
The 3000 windings are directly wounded onto the mounting using thinned[20] GeVarnish[21] to fix the windings. It is the standard glue for winding coils providing a good thermal contact. Nevertheless it is known that it aect the FormVar [22] insulation on the wire. To circumvent electric shorts between the Cu-matrix and the mounting, a thin mylar[23] is used between the windings and the Cu-mounting, at all types of mountings, as mechanical stress can cause the inner windings to be damaged A.3.1. A picture of one vertical transport coil wounded onto a Cu-mounting is shown in Fig.(13.7).

---
[18]GE Low Temperature Varnish, from Oxford Instruments, Abingdon Oxfordshire OX13 5QX, Great Britain: find material properties in the Appendix.
[19]54S43, Wire size: 0.10mm bare, 0.127 mm diameter with FormVar insulation, serialNo. 378E-98B3B2A, from Supercon Inc. 830 Boston Turnpike, Shrewsbury, MA 01545
[20]GeVarnish is thinned (1:1) with Ethanol to keep the glue during the winding process solvent.
[21]GE Low Temperature Varnish, from Oxford Instruments, Abingdon Oxfordshire OX13 5QX, Great Britain
[22]FormVar is used as the insulation at the SC wire.
[23]0.025 mm thick mylar ICEles40 from ICE Oxford: find material properties in Appendix D

13.4. The superconducting coils

The standard coil mounting as it was originally designed is the B-type mounting with cap, to ensure thermal contact from almost three sides. As all mounting-types it is sliced into four segments and glued together with Stycast[24], to maintain mechanical stability while preventing a closed loop over the circumference.
The C- and D-type mountings are without cap, to reduce the dimension of the mounting as it is needed for the lowest coil[25] and the upper coil[26]. Putting a coil on such a modified mounting, two ways are possible either directly winding the coil an the mounting (C-type mounting) or winding a self-sustaining coil on a Teflon mounting, and than attaching it onto the Cu-mounting, using again mylar and GeVarnish (D-type mounting).
Tab.(13.2) therefore shows the dierent coil- and mounting types in the four transport slots, each at a mean distance of 30mm. Fig.(A.5) shows a B-type mounting with the coil directly wounded onto the Cu-mounting, inset a), and the way to achieve a D-type mounting with a self-sustaining coil in inset b)-d).

In dierence to the high current room-temperature coils, the sc-transport coils provides currents up to 1.5A, with a total resistance of ≈ 0.5 maximum, as just the normal conducting parts contribute to the resistance. Therefore the power supplies do not have to provide huge currents or voltages, but must handle the large inductance of the coil which results from the 3000 windings. In addition the currents in the coils must be switched in both directions for the vertical transport, raising the question for a smart way to control the coils.
It turned out that a single power supply with an additional H-bridge as applied in the normal-conducting vertical transport section can not handle the fast switching of the coils which is demanded[27]. This was easily solved using two current sources[28] operating counter-wise, and triggered via a 5A-switch[29]. This fast acting switch therefore provides the protection for the coils, as the high induction voltage during switch-o has to be considered. Hence an implemented quenching circuit as shown in Appendix (B), cancel the current through the switch, while simultaneously limiting the voltage at the coil.

13.4.2. The QUIC- and super-Ioe-coils

The QUIC-coil is located 19mm from the transport axis in respect to the nearest windings. It sustains 1800 windings with an inner coil radius of $R_i = 5mm$ and and outer diameter of $d = 20mm$. The coil has a length of 8mm and is mounted on a Cu-mounting, the Ioe-mounting which is sliced in two halves to again prevent a closed metal loop susceptible to eddy-currents. It is directly wounded onto the mounting, and glued with GeVarnish, to maintain thermal contact between the windings. Fig.(13.7), inset b) shows how the ioe-coils are fixed on the Cu-mounting. To protected them from thermal radiation, as they are located rather close to the windows, Al-tape is carefully fixed on all bare windings whereas again a closed Al-loop is avoided (not shown in Fig.13.7).
The super-ioe coil directly glued on top of the mounting, inset b) consists of 250 windings

---
[24]Stycast 2850FT/11: from Emerson and Cuming
[25]This coil is closest to the bottom of the thermal shield, and should not contact the aluminum at 55K.
[26]As the Ioe-mounting has to fit in between the uppest coils, the cap of coil V_9 is removed, that the mountings fits in
[27]Protection-diodes in the H-bridge design would prevent the coils to follow the driven currents
[28]For a detailed scheme of the experimental current control for the sc-coils see Appendix (C)
[29]For more details see Appendix B

13. Setup of a 4K cryo-system

and has a length of 2mm. Due to the rather bad thermal contact, this coil is capable of maximum 350mA whereas the QUIC-coil can sustain currents up to 1.35A.

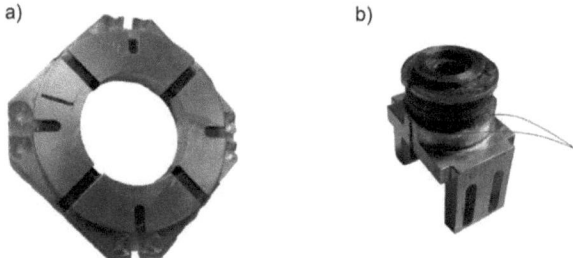

Figure 13.7.: Inset a) shows a vertical transport coils wounded onto a Cu-mounting, which is sliced to prevent eddy-currents. Inset b) shows the Ioe- and the super-Ioe-coil on the corresponding Cu-mounting. In addition both coils are protected from thermal radiation using Al-tape carefully wounded around the bare windings

To implement a serial circuit for the QUIC-trap, and therefore increase the stability of the trap, an adequate current control has to be implemented. As the two coils V_8 an V_9 participate in the transport and are controlled independently during transport, the Ioe-coil has to be turned on in the later phase of the experiment, possibly using the same current as in the quadrupole trap at the end of the vertical transport. This is achieved using a power-splitter as shown in Appendix (C).

13.4.3. Solder contacts for the SC coils

Al superconducting coils are connected to the normal-conducing wires on the base-plate via solder contacts. It turned out that the most reliable way for maintaining a low dissipative electric contact is, to solder both ends of the normal- and the super-conducting wire into a small Cu-tube using a lead-containing solder[30]. Additionally to this approach, several dierent methods were tried out, all facing dierent drawbacks and limiting the maximum coil currents well below the desired values above 1A (see Appendix A.3.2). Goldened Cu-bobbins are used to thermally anchor the wire ends next to the contact. Nevertheless this is important as the dissipated heat in the solder contact must not heat the fragile superconducting wires anyway.

13.5. 4K system and the transport section

The complete experimental setup relies on the two chambers, the MOT/preparation-chamber and the cryo-vacuum-chamber with the cold finger. Connected via the transport section, it is quite of interest which pressures in the system can be achieved, as the MOT chamber was

---
[30]Solder of type: HF32, S-Sn60Pb39Cu1

13.5. 4K system and the transport section

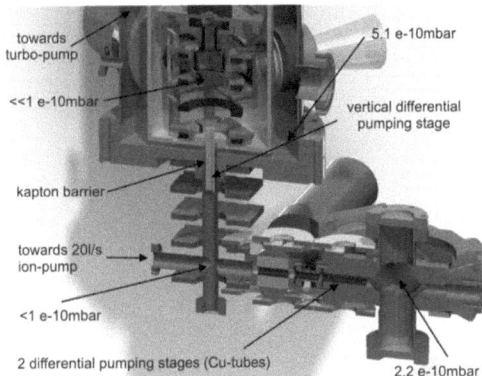

Figure 13.8: Surprisingly low pressure is achieved in the outer cryostat-vacuum chamber as this region is now strongly decoupled from the transport section and the inner 4K part. This is done by closing most of the holes in the upper part of the thermal shield, and by introducing Kapton-rings around the Al-differential pumping stage at the intersection of transport and cryostat. These rings even enhance the differential pumping capability of the Al-tube, nevertheless pumping the transport section becomes more difficult in the early stage of evacuation, while the ion-pump in the transport section is still of.

carefully prepared for UHV, and the cryo-section with its implementations is quite a *dirty* environment, nevertheless operated at very low temperatures.
Fig.(13.8) shows a cut through the experimental setup. The differential pumping stages (Cu-tubes) in the horizontal transport section and the Al-tube in the vertical transport section therefore allows to reduce the pressure in the corner of the transport to values below 1×10^{-10}mbar. To completely separate the outer vacuum chamber and the transport section a Kapton barrier[31] was used. The pressure in the outer cryo-chamber is surprisingly low, as the chamber was not well prepared for UHV[32], keeping in mind, that the thermal radiation shield operated at 50K supports vacuum as it cryo-pumps the outer chamber. Nevertheless, it turned out that the connection holes in the upper top of the thermal shield are a drawback, as then the pressure in the 4K environment is affected by the pressure in the outer vacuum chamber (see Appendix A.5 for further pressure issue).
To maintain as good pressure as possible in the 4K environment, the region is partially decoupled to the transport section via the Al-tube, the differential pumping stage at the bottom shield, as well to the outer vacuum chamber via the Kapton barrier and by closing most of the holes in the top shield cap. In addition three char-coal container support the 4K cryo-surface to allow for pressures well below 1×10^{-10}mbar.

To support the evacuation process, heating stripes in the vertical transport section allows for temperatures up to $200°C$. Surprisingly, the cryo-finger reaches temperatures slightly below the steady-state values of 5.1K base-temperature and 52K shield-temperature, while the transport section is baked out. After reaching steady-state, the heating stripes are turned

---
[31]The Kapton barrier (3 rings of Kapton-foil) was introduced to separate the outer cryo-chamber from the transport and the 4K environment
[32]As the chamber is opened quite often for rebuilding the cryo-finger, it should be baked out every time, considering that then the inner 4K with the helium must be removed. As this is too complicated doing it every time, the chamber is not treated anyway.

13. Setup of a 4K cryo-system

o, and the vacuum in the outer cryo-chamber drops down to $5 \cdot 1 \times 10^{-10}$mbar.
Fig.(13.9) gives an overview how the experiment looks like. The breadboard, inset a) which maintains the imaging optics for the cryo-chamber introduces a lower and an upper setup-layer, with the MOT-chamber and the transport below it, inset b). Inset c) and d) show a closeup of the vertical transport section, with the two chambers disconnected, and connected with adjusted vertical transport coils. Inset c) clearly shows how the uppest transport coil V_5 with its water-cooling is directly attached to the clearance in the CF200 bottom flange.

Figure 13.9.: Overview of the assembled setup for transport of atoms into a cryogenic environment. Inset a) and b) show the two-layer setup with the lower-magnetic chamber below the breadboard and the cryogenic system above. Inset c) and d) show a closeup of the vertical transport section connected to the bottom flange of the cryostat chamber

13.5.1. Upper imaging system

To image the atoms in the cryo-system, absorption imaging (see Appendix A.8) is applied. Laser light from an out-coupler goes along one optical axis of the 4K system and is captured by a high performance camera[33]. The CCD covers $1024px \times 1392px$ with a pixel size of $6.45\mu m \times 6.45\mu m$. The objective in the focal distance of 243mm provides a two-fold magnification. The imaged area in the trapping region therefore is $\approx 13mm \times 17.6mm$ which is fairly enough to image the atoms in the quadrupole- and the QUIC-trap[34].

---

[33] *Pixelfly qe* camera from pco.imaging
[34] The QUIC-trap might be shifted up to 7mm from the trapping-center.

14 Results: Cold atoms inside a 4K-cryogenic machine

14.1. Transport of atoms into the cryostat

The atoms are trapped and laser-cooled in the MOT chamber and then moved into the cryostat, and trapped in the quadrupole trap following the sequence shown in Tab.(11.2). The atoms are first horizontally transported into the corner, and then vertically moved up into the final trapping position at $z = 215mm$ in between the uppest transport coils.

14.1.1. SC vertical transport scheme

Dierent transport parameter (see section 10.3.2) allow for flexible transport schemes adjusting the transport time, velocity and acceleration. As for the horizontal part the transport velocity is given by $x_{final} = 210.4mm$ and transport time T, where the maximum velocity can be arbitrarily defined to be reached at $T_{max} < T$, the vertical transport works dierent. The superconducting coils with its large inductance are limiting switching times. Therefore the maximum acceleration a_{vert} must be limited instead of being defined by the time of reaching maximum velocity T_{max}, as for the horizontal transport. With an additionally defined vertical transport velocity v_{vert}, the vertical transport time is given and can mainly be influenced by the velocity if acceleration and deceleration are done quite fast ($\approx 1 - 4m/s^2$).

In principle one wants to transport the atoms as fast as possible, to loose less atoms during transport, as the pressure in the transport section is higher than in the cryogenic environment. For experimental reasons it also seems desirable if the overall cycle-time can be kept short. It turned out, that with a maximum acceleration of $4m/s^2$ and a velocity of $0.15m/s$ atoms can quite eciently be transported into the cryo, as the transport currents follow still the preset currents within accuracy of $\approx 5\%$ at the peak-value. The maximum deviation $\Delta I_{max} = I_{set} - I_{coil}$ of the vertical superconducting transport currents can be found in Tab.(14.1), where the positive deviations depict, that the current source could not completely follow the preset currents[1].

---
[1] The negative deviations occur due to bad calibration of the current driver.

14. Results: Cold atoms inside a 4K-cryogenic machine

sc-coil	current source for	ΔI_{max}	nc-coil	ΔI_{max}
V_6	I6a	5.7%	V_1	7.6%
V_6	I6b	5.6%	V_2	3.4%
V_7	I7a	5.0%	V_3	5.2%
V_7	I7b	−3.3%	V_4	4.4%
V_8	I8a	4.3%	V_5	3.3%
V_8	I8b	−1.3%		
V_9	I9	4.2%		

Table 14.1.: Deviation from preset transport currents for both, superconducting (sc) and normal (nc) conducting vertical transport coils. For positive sign the values show that $I_{measure}$ was smaller than the preset current, for negative sign, the deviation is mostly due to a non-constant calibration of the current clamp (Type:).

As counter-operating current sources are applied at the superconducting transport coils, the positive and the negative half wave of transport currents are labeled a and b for a single coil, where I_{6a} describes the current through coil V_6 if the atoms are below the coil center and and I_{6b} if the atoms are already towards coil V_7.

14.1.2. Imaging the atoms in the cryostat

Applying a time-of-flight (TOF)[2] after switching of the trapping fields and shining in a resonant, or near resonant laser-beam on the hyperfine transition $5^2S_{1/2}|F=2\rangle \rightarrow 5^2P_{3/2}|F'=3\rangle$, allows to observe the expansion of the atom cloud during free fall. The atom number can easily be derived from the cloud size and density (see Appendix A.8), whereas the temperature as well as the lifetime is derived from TOF-measurements. Keeping the trap on, allows to take an in-situ image of the atoms, whereas most of the atoms are detuned from the imaging transition as they feel the trapping potential which shifts the Zeeman-split hyperfine manifold space-dependent.

Fig.(14.1) shows two insets with the atom cloud imaged at resonance and in-situ with $TOF = 0ms$ inset a), and at $TOF = 6ms$, inset b), both after a hold time of $t_{hold} = 5ms$ between final transport and switching o the trap. The graph shows a completely non-expected, systematically oscillating absorption signal of atoms with accumulating TOF. This observation was extended by scanning both, TOF and detuning ( ), as shown in Fig.(14.2), which shows such a 3D detuning scan and the corresponding atom number in the lower inset for = $0MHz$. The 3D scan indicates how the vanishing absorption signal between $2.2ms < TOF < 3.5ms$ in Fig.(14.1) can be explained. As the magnetic quadrupole trap is switched o to release the atoms, an oscillating magnetic field is obviously induced, which shifts the Zeeman-sublevels of the imaging transition as for atoms trapped in $|F=2, m_F=2\rangle$, $\mu_B/\hbar = 1.4MHz/G$. The atoms are therefore even after some milli-seconds up to 30MHz blue-detuned from the imaging transition and as the magnetic field decreases, come more and more in resonance.
An additional rotation of the magnetic field vector would also change the polarization axis of

---
[2]Find more details to the time-of-flight measurements in Appendix A.8.

14.1. Transport of atoms into the cryostat

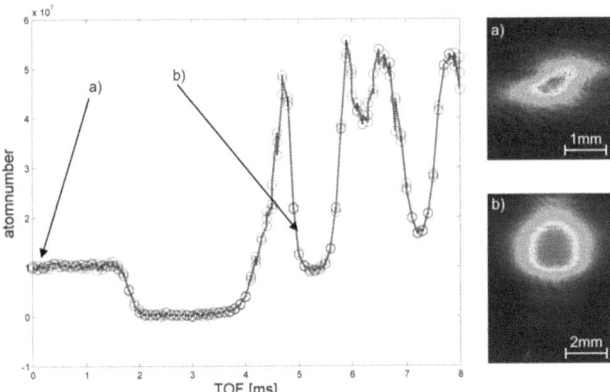

Figure 14.1.: TOF scan of atoms trapped in the superconducting quadrupole trap. Inset a) shows an in-situ absorption image after TOF=0ms at resonance, and inset b) shows an absorption image after TOF=6ms at resonance, both after a hold time of 5ms in between transport and TOF

the imaging beam in respect to the atoms. Thus the Clebsch-Gordan coefficients [3] of different m_F-sublevels are not equal, and the absorption cross section would change, hence influencing the measured atom-number in the 3D-scan, Fig.(14.2) as described in Appendix (A.8). All observations indicates that this *revival of the absorption signal* is due to an oscillating magnetic field, additionally influenced by eddy-currents which would cause a damping of the oscillations. This oscillating magnetic field could therefore be caused by a resonance-effect between different super-conducting coils, non-simultaneous quenching of the trapping currents, and additional eddy-currents. It is in detail described in Appendix (A.6).

14.1.3. Spatial oscillations during free fall

An oscillation of the magnetic field after switch off can be measured either by a 3D detuning-TOF scan Fig.(14.2), or by monitoring the induction voltage Fig.(A.9) at a sc-coil as shown in Appendix (A.6).
The oscillating magnetic fields therefore influences the free expansion of the cloud and temperature measurements due to TOF imaging becomes difficult.

An evidence that the oscillations shown, results in field gradient which occurs and vanishes, is found by the comparison between the measured and the calculated cloud position after *free fall* for a certain TOF. The calculated position therefore reads as:

---
[3]Find CGK for rubidium D2 line in [140]

14. Results: Cold atoms inside a 4K-cryogenic machine

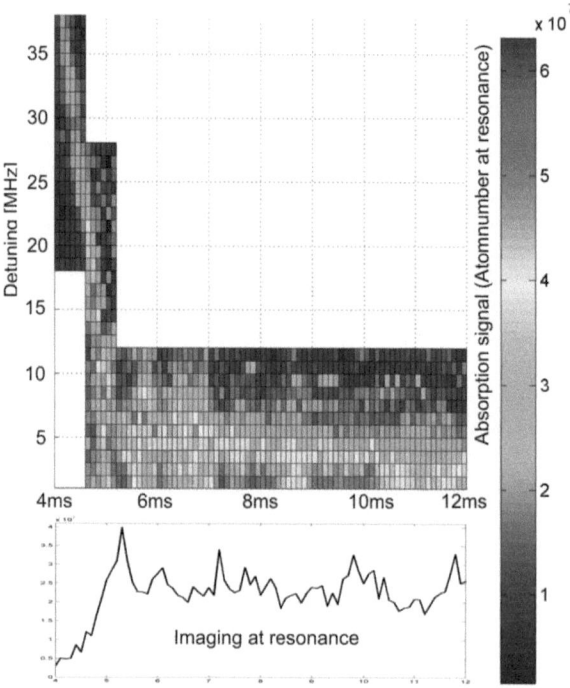

Figure 14.2.: Oscillating atom number with detuning in a TOF-scan. The revival of the absorption image even at dierent detunings indicates, that a damped magnetic field oscillates at the position of the trapped atoms even after switching o the trapping currents

$$\Delta s = \frac{g}{2}\left[TOF_1^2 - TOF_0^2\right] \tag{14.1}$$

with $TOF_0 = 4.5ms$ according to Fig.(14.3). The lower graph shows a comparison for the calculated position according to Eq.(14.1) after releasing the atoms from the trap, and the measured vertical position. Indicated by the upper inset, which shows an oscillating absorption signal as shown in Fig.(A.9) at zero imaging detuning, the deviation between calculated and measured cloud position is minimum for maximum absorption signal and therefore nearly zero magnetic field, or even a vanished trapping potential. For lower atom number, and therefore an indicated higher detuning due to a present magnetic field, the atoms are pulled back and upwards towards the original trapping position.

14.2. Optimization of the transport

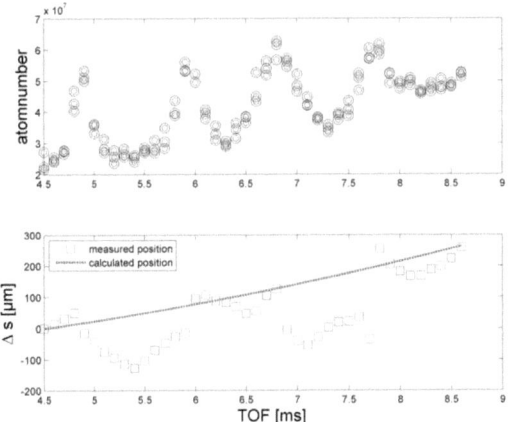

Figure 14.3.: The upper inset shows the same measurement as shown in Fig.(A.9), the oscillation of the absorption signal for zero-detuning as the atoms are Zeeman-shifted and imaging polarization rotates. The lower graph shows the corresponding vertical position for a certain TOF which indicates that for maximum atom-number, the trapping field is nearly to zero, which leads to an almost perfect correlation between calculated and measured trap position.

14.2. Optimization of the transport

14.2.1. Lower magnetic trap

Loading the magnetic trap in the MOT chamber follows two simple demands. First, loading as much atoms into the trap as possible and second, with a temperature low enough to allow for an efficient transport. The MOT was optimized and characterized in terms of cooling-light power, dispenser current, MOT-field current and loading time. Find further details in [178]. It is capable of preparing about 10^9 atoms for the optical molasses with temperatures of $40\mu K$ before reloading and 5×10^8 atoms in the magnetic trap with a temperature of $150\mu K$ respectively.

Heating rate

It turned out, that the lifetime of atoms in the magnetic trap is rather poor, and with $= 750ms$ well below several seconds. The magnetic trap is stepwise turned on from 0 to $22G/cm$ in $50\mu s$, than within $0.5ms$ to an intermediate trap with $50G/cm$ and finally to $130G/cm$, totally within 6ms, as longer ramping times did not show an improvement of the heating rate, or the loading efficiency. Due to the short lifetime it could not be figured out if the temperature increase from $40\mu K$ in the optical molasses to $150\mu K$ in the magnetic trap

143

14. Results: Cold atoms inside a 4K-cryogenic machine

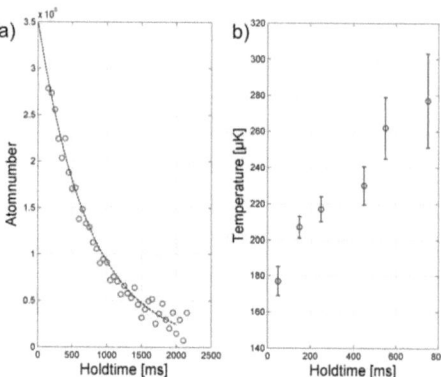

Figure 14.4: Lifetime and heating rate in the lower magnetic trap. The characteristic time constant of $\approx 750ms$ indicates that the background pressure is rather high in the MOT-chamber, which is indeed true ($p \approx 5 \times 10^{-10} mbar$). To start the transport of atoms with $T < 200\mu K$, the holding time before transport should not last too long, as otherwise heating due to background collisions would cause heating of the trapped atoms.

could be lowered by turning the trap on more slowly. Both reasons, the short lifetime in the magnetic trap, and the heating rate of $\approx 14\mu K/100ms$ suggests to bring the atoms as fast as possible out of the MOT-chamber, as the atom temperature must be kept in the order or below $200\mu K$ at least, to ensure a working transport. Fig.(14.4), inset a) shows the exponential decay of the atoms with holding time, and a characteristic time-constant = $750ms$, while inset b) shows the temperature of the atoms in the trap, rising from $150\mu K$ up to $280\mu K$ within 700ms.

14.2.2. Horizontal transport

After rebuilding the transport and dismantling the horizontal transport coils, as well as coupling the transport tube to the cryostat, a final transport e ciency of more than $_{hor} = 90\%$ in one direction was found after careful adjustment of the transport coils. As a previous characterization [178] showed even without opimization results up to $_{hor} = 80\%$, this was easily overcome by applying the optimal transport parameter. It turned out, that even due to the limited lifetime in the MOT-chamber the optimal position to reach full transport speed is 150ms after launch. This results in a fast acceleration out of the chamber, whereas the atoms are than for the rest of the 550ms transport duration smoothly shifted towards the final horizontal position. The atoms show a moderate heating after the transport in the order of $50\mu K$[4].

The e ciency is measured by bringing the atoms into the corner and back, while comparing the initial and final atom number to the corresponding number after 1120ms hold time in the chamber. Even if this method does not account for di erent losses due to collisions with background gas as the pressure along the transport section might change, it gives a good

---
[4]This increase in temperature might even for the one-way transport be smaller as the atoms are transported back and forth to be characterized in the lower chamber. The additional accelerations and decelerations of the trap might induce additional parametric heating in the trap

14.2. Optimization of the transport

feeling for the overall efficiency.
Fig.(14.5) shows the influence of the fast acceleration out of the MOT chamber. The blue and the red curves give two independent measurements, before (red) and after (blue) aligning the horizontal transport coils. The stepwise decrease in efficiency after $x = 40mm$ was overcome by accelerating the atoms faster as shown in inset b), whereas inset a) shows a uniform acceleration over the transport distance. A fast acceleration seemed to prevent the atoms to be lost after $x = 160mm$, as the time duration where the aspect-ratio[5] of the trap opens can be kept short.

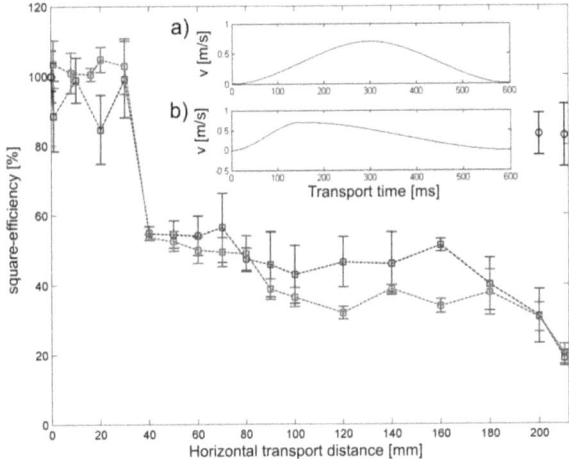

Figure 14.5.: Horizontal transport efficiency even in the region $x \approx 35mm$ was increased by accelerating the atoms rather fast out of the chamber, compared to the measured curves (red,blue) in two independent runs, before and after adjusting the horizontal transport coils. Inset a) and b) exhibits the different acceleration schemes, where the application of scheme b) even lead to a horizontal transport efficiency of more than 90%, and which finally overcame the big loss in efficiency at $x \approx 35mm$. The efficiency-square was therefore just meausured for the complete horizontal length.

14.2.3. Vertical transport

The following describes why a vertical transport characterization, namely the determination of the efficiency by forth-and-back transport as used in the horizontal transport, can not be applied.

---
[5]See section (10.3) were the opening of the aspect ratio is described.

14. Results: Cold atoms inside a 4K-cryogenic machine

The long lasting forth-and-back transport of the atoms into the cryo-chamber and back to the MOT-chamber, lasting more than 5 seconds, shows that the pressure in the horizontal transport line, even in the MOT chamber is a limiting factor. The measurement of finally transported atoms is compared to atoms remaining after the same time $t_{hold} = t_{transport}$ in the lower chamber, and thus results in a transport e ciency $_{total} > 100\%$, which is obvious as the pressure in the lower MOT chamber is by all means higher than in the transport section and the cryogenic environment. Hence the vertical part is characterized by just optimizing the number of trapped atoms in the cryogenic environment.

It is always a trade-o between transport time and acceleration of the magnetic trap, transporting as much atoms as possible into the cryogenic environment. Fig.(14.6) therefore shows a 3D-plot of trapped atoms, imaged after a TOF=5ms at zero imaging detuning.

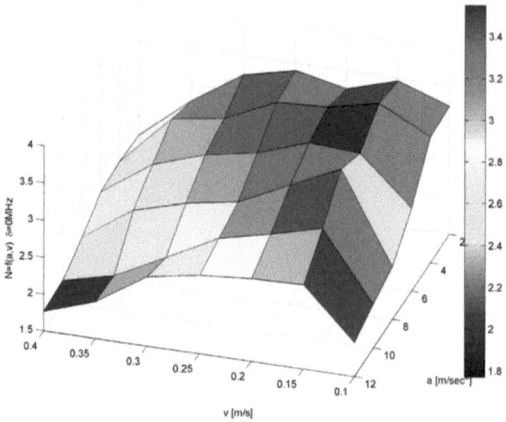

Figure 14.6.: Atoms imaged at resonance after a 5ms lasting hold-time in the sc-quadrupole trap depending on vertical transport parameter

Facing a measurement uncertainty of $\approx 10\%$ and a measurment drift in the same order, even the best transport parameter can be found at $v_{vert} = 0.15m/s$ and $a_{vert} = 2m/s^2$ as shown in Fig.(14.6).

The long lasting transport time, mainly defined by a weak acceleration, is clearly a limiting factor as shown in Fig.(14.7), inset a), whereas above $a_{vert} = 4m/s^2$ the current source can not follow the set current. Inset b) shows therefore the drop in trapped atoms for weak accelerations below $1m/s^2$.

14.3. SC quadrupole trap

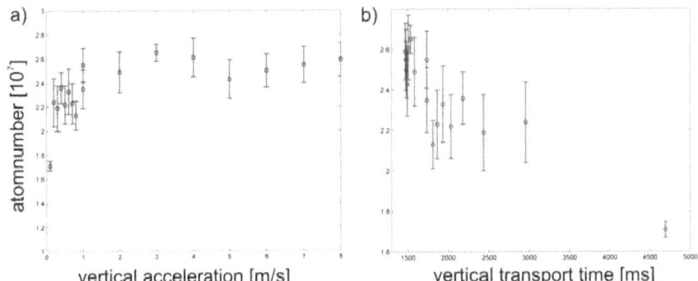

Figure 14.7.: Number of trapped atoms for an optimal vertical transport velocity of $v_{vert} = 0.15$. For small acceleration in the order of $0.1 - 0.2 m/s^2$, the transport time require almost 5s

14.3. SC quadrupole trap

As for the first time on 21^{st}September 2010 a cloud of $\approx 1.2 \times 10^6$ atoms was magnetically transported into a cryogenic environment, the maximum atom number was since than optimized to 7.0×10^7, which is not the absolute maximum, as the applied absorption imaging underestimates the number of trapped atoms (see Appendix A.8).

14.3.1. A large superconducting quadrupole trap

It was figured out that even a trade o between the initial trap temperature and the number of trapped atoms in the lower magnetic trap before transport, can lead to a higher atom number in the cryostat, as shown in Fig.(14.8). Inset a) shows the 2D scans of atom number (upper 3 pictures) and the root-mean square (RMS) of the the Gaussian density distribution (lower row) vs. molasses hold-time and molasses detuning for dierent time of flights. As even in the upper left corner for 2ms hold time and $= -70 MHz$ the RMS value[6] and the atom number is lowest, more and warmer atoms can be trapped in the initial magnetic trap, as shown in inset c).

The slight temperature increase as shown in Tab.(14.2), does not much aect the transport much, as even at a higher detuning of $= -30 MHz$, the atom number in the cryo drops, inset b). A maximum of atoms transported into the cryo can be found at $= -40 MHz$, Fig.(14.8). Nevertheless this rises the question, how the final atom-temperature in the superconducting quadrupole trap is influenced by the higher initial transport temperature.

---
[6]For a algorithm to calculated the temperature from the RMS refer to Appendix A.8

14. Results: Cold atoms inside a 4K-cryogenic machine

mol [MHz]	t_{hold} [ms]	T_{trap} [μK]	N_{trap} [atoms]
-70	10	179.6± 4.1	3.48± 0.25 10^8
-70	2	178.0± 5.0	3.47± 0.24 10^8
-40	10	207.4± 4.7	4.15± 0.20 10^8
-40	2	205.5± 5.5	3.87± 0.27 10^8

Table 14.2.: Temperature and atom number in the initial magnetic trap which indicates that even for lower detunings at $= -40MHz$ more atoms can be transported if the slightly increased temperature does not affect the transport much.

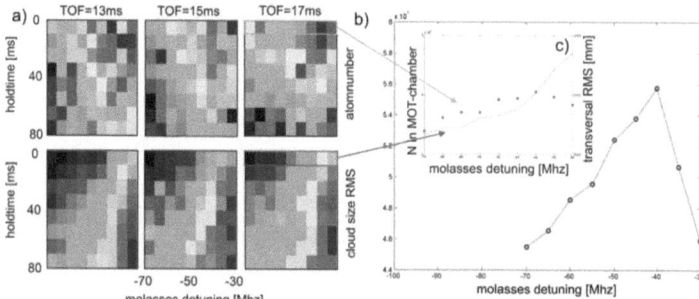

Figure 14.8.: Inset a) shows a series of TOF scans for molasses detuning and molasses hold time for both the RMS value of the Gaussian density distribution and the atom number in the lower magnetic trap. Inset c) therefore shows a line-scan through the TOF pictures for $t_{hold} = 10ms$, and inset c) gives shows that even with a lower molasses detuning more atoms can be transported into the cryostat

14.3.2. Lifetime in the cryogenic environment

The lifetime in the quadrupole trap was stepwise improved from initially $_{life} = 4 - 5s$ to well above $_{life} = 200s$ as shown in Fig.(14.9). This was maintained by different provisions, namely by improving all different pumping stages from the cryogenic environment to the outside and by implementing char-coal container (see section 13.5) at the 4K environment.

Fig.(14.9) exhibits a lifetime of $223 \pm 41s$ for the atoms in the superconducting quadrupole trap. As the initial transport temperature is relatively high, in the order of $200 \mu K$, the atoms are even after transport quite hot and are not undergoing *Majorana* spin-flips. Among all other loss mechanisms, excluding light assisted collisions, the decay rate of atoms can be fitted according to a single exponential decay ($N = N_0 e^{-t/\ _{life}}$) as the hottest atoms after transport are lost quite fast (refer to 5.4). The complete discussion about limiting factors on the lifetime of the atoms can be found in Appendix (A.5).

14.3. SC quadrupole trap

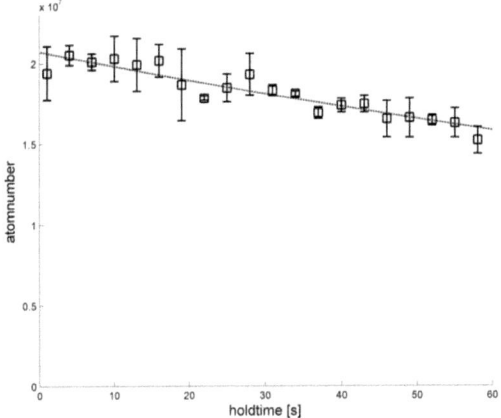

Figure 14.9.: Lifetime of atoms in the superconducting quadrupole trap

14.3.3. In-situ tomography of the sc-quadrupole trap

Since the imaging beam can easily be detuned into the blue, it is trivial to do an in-situ tomography scan of the atoms trapped in the superconducting quadrupole trap. Since this allows a deeper insight in the trap geometry maintained by the superconducting coils, and the size of the atom-trap, this seems quite feasible. Fig.(14.10), displays such a scan where insets a)-l) show for dierent detunings the atom density. The atoms are projected in the imaging direction on iso-B-field surfaces. Bright spots depict regions with higher atom-density, nevertheless an imaging eect comes into play. As described in the *Method-section* (A.8), a well defined imaging process demands a well directed magnetic guiding field to define a quantization axis regarding to the polarization of the incoming laser-beam. As this is not provided in this measurement, atoms exhibit dierent *Clebsch-Gordan* coecients at dierent positions, as the quadrupole field-lines do not provide a constant quantization axis. Therefore the scattering-cross sections are dierent and areas with same atom-density may occur brighter or darker. This can clearly be seen in insets g)-l) were maximum atom number occurs at the outer edge of the cloud.

Even for reloading the atoms onto a chip-based trap (see section 15.1), it is useful to know the trap-size and the trapping gradients of the macroscopic magnetic trap in detail. For a detuning of $= +29.2$MHz the absorption picture therefore even exhibits atoms at an iso-B-field line of $|B| = /\mu_B$ with a vertical dimension of the cloud which is $d_V \approx 3mm$. For atoms trapped in $|F = 2, m_F = 2\rangle$ the Lande-g factor and the quantum number gives $m_F \cdot g_F = 1$, and with $\mu B = 1.4$MHz/G therefore resulting in $|B| = 20.9$G, due to the *Zeeman-shift* of the hyperfine levels. Hence, the vertical trapping gradient can be estimated

149

14. Results: Cold atoms inside a 4K-cryogenic machine

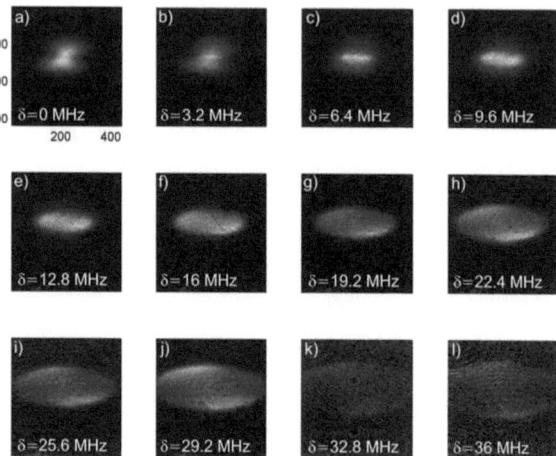

Figure 14.10.: In-situ images for various blue-detuned imaging light. The brightness corresponds to higher atom-density in the superconducting quadrupole trap. The plots represents projections along the imaging axis for atoms on iso-B-field lines, and exhibit the size of the magnetic trap. The axis are given in pixels for a size of $12.65 \mu m/px$.

as $B'_z = \frac{2|B|}{d_V}|_{=29.2MHz} \approx 139 G/cm$. In fact this tomography method can be used to validate the results shown in Fig.(A.8), inset d), were the vertical magnetic gradient for real coil geometries with the implemented shaped transport current I_6 is calculated to be $\approx 140 G/cm$. As the trap gradient is rather high, the gravitational-sag of $15 G/cm$ can be neglected.

14.4. A superconducting Ioffe-Pritchard-like trap

Following section (5.3.3), the measured position shift of the trap can be compared to the calculations. As the magnetic transport is ending, the Ioffe-current is ramped up, therefore shifting the position of the atom-cloud towards the QUIC-coil. In the region of $\Delta x \approx 6.5 mm$, the trap bottom is lifted from zero, maintaining a superconducting QUIC-trap. Fig.(14.11) exhibits the difference in measurement and calculation. The blue curves are calculated trap shifts with either a coarse geometry (o), or a fine geometry (◇) chosen for the number of windings of the coil[7]. The calculations are done for $I_{quad} = 400 mA$ and $I_{Ioffe} = 500 mA$, while the measurement (o) was also performed at $I_{quad} = 400 mA$. In addition the right axis exhibits the trap bottom, calculated for a fine-geometry. For the chosen parameter, the calculation results in $\Delta x \approx 6.455 mm$, at $B_0 = 3.3 G$ with $_\parallel \approx 2 \times 15 Hz$, a rather weak and

---
[7]Fine-geometry: $N_{quad} = 7 \times 3$, $N_{QUIC} = 5 \times 8$ windings, coarse-geometry: $N_{quad} = 21 \times 9$, $N_{QUIC} = 32 \times 56$ windings, both with the nominal current density and the nominal dimensions as shown in section (13.4)

14.4. A superconducting Ioe-Pritchard-like trap

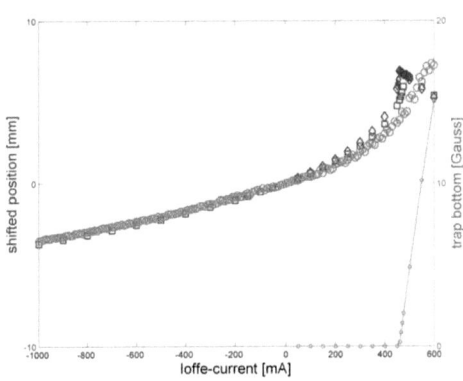

Figure 14.11:
Comparsion of measured and calculated position shift of the superconducting quadrupole trap. The red curve (o) depicts the measurement, whereas the blue curve shows the calculation for a setup realized in the experiment with $I_{quad} = 400mA$ and $I_{Ioffe} = 500mA$, for both, coarse (o) and fine-resolution (◊) regarding the calculation of $M \times N$ windings of the superconductive coils. The trap bottom of the QUIC-trap for higher Ioe-currents is calculated for the fine-resolution and shown in green.

macroscopic QUIC-trap.

As the simulation indicates that for $I_{quad} = 400mA$ and $I_{Ioffe} = 500mA$ the trap bottom is even at $B_0 = 3.3G$, a measurement of the lifetime exhibits a further attribute of the QUIC-trap. After establishing the QUIC-trap, the trap bottom depends strongly on the Ioe-current and thus introducing parametric heating [8]. This would in addition cause the atoms to be lost, as the hottest atoms immediately drops out of the trap, and an appreciable decrease in atom-density. This process would occur on a much faster timescale than the losses due to inter-species- or background-gas-collisions. Fig.(14.12) shows such a lifetime measurement indicating that there is huge noise on the Ioe-current, respectively a huge instability[9] between I_{quad} and I_{Ioffe}. According to the loss rates in low density atomic clouds Eq.(5.24) and [91], the decrease in atom number can be fitted with

$$N(t) = N_0 \ e^{-\frac{t}{\tau}} \frac{1}{1 + {}_0N_{\ 0}[1 - e^{-\frac{t}{\tau}}]} \tag{14.2}$$

were $= 1/$ is proportional to the losses through background pressure. Inelastic two-body losses as well as heating therefore enters in ${}_0$. The fitting parameter were determined as $N_0 = 9.51 \pm 0.31 \times 10^6$, $= 239.5 \pm 68s$ and ${}_0 = 1.98 \pm 0.02$ with 95% confidence.

The lifetime in the QUIC-trap was measured after a previous lifetime-measurement performed in the quadrupole trap, indicating a lifetime well above 100s. Inset b) of Fig.(14.12)

---
[8]For parametric heating see section (5.4).
[9]When this measurement was performed, the serial power-splitter (Appendix C.1) was not implemented. The power-splitter allows to define a controllable ratio between the three currents I_{8b}, I_9 and I_{Ioffe} using partially driven MOSFETs, to connect the three coils in series driven by just one power supply, and hence increase the relative stability between the currents.

14. Results: Cold atoms inside a 4K-cryogenic machine

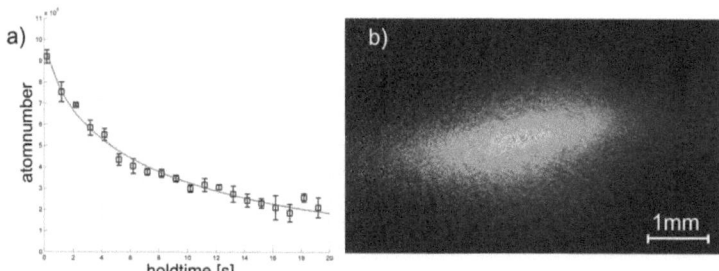

Figure 14.12.: Lifetime of atoms in the QUIC-trap for $I_{quad} = 400mA$ and $I_{Ioffe} = 500mA$, imaged at resonance after $TOF = 6.5ms$. The exponential decay indicates that parametric heating causes the hottest atoms to be lost which results in strong decrease in trap-density at the bginning of the measurement. This can be regarded to relative variations between the Ioe-current and the quad-currents which causes instabilities and therefore heating.

shows a picture of the atoms released from the QUIC-trap after $t_{hold} = 100ms$ and a TOF of 6.5ms, imaged at resonance.

15 Towards: Cold atoms in a superconducting micro-trap

Reloading the atoms onto a superconducting, chip based micro-trap, will follow the scheme described in this chapter, where also the underlying calculations to achieve adiabatic reloading without switching o the trap completely at any time, are presented. Furthermore, a preliminary chip-design is shown with a chip-mounting capable of implementing microwave lines, as the experimental setup should in principle allow for magnetic coupling of cold atoms to microwave fields.

A sketch of the reloading setup is shown in Fig.(15.1) where a transparent chip surface with a Z-shaped wire is illustrated. In principle the chip just demands the Z-structure for reloading and comes without a huge closed superconducting surface. The trapping wire will be made either of Nb or NbTi, a thin film, sputtered onto the substrate.

Considering this setup, the *Meissner-e ect* is less an issue for reloading than for trapping the atoms close to the surface as it is known that the current-density distribution in superconducting wires changes dramatically [103, 104, 105], and hence influencing the trapping potentials in respect to the conventional room-temperature wire traps, if the trapping distance is in the order of the wire dimensions. Nevertheless the non-existent superconducting surface would prevent this setup from vortices and upcoming vortex-traps [201], originating from perpendicular field-components penetrating large superconducting surfaces.

15.1. Re-loading atoms in a superconducting micro-trap

Starting point for the reloading is the QUIC-trap (see section 5.3.3), as this provides two major advantages: A) The atoms can be pre-cooled in a trap with a non-zero trap-bottom using evaporative cooling [202], and the reloading can be done adiabatic without switching o the trap completely [1]. The trap can even be more compressed increasing the phase space density during the cooling process as *Majorana-flips* are prevented (see section 5.4).

---
[1] Since the Io e-like chip-trap has still a quadrupole shaped trapping geometry tilted by 45 ° in respect to the transport trap, the atoms would be lost, loading them from a zero-bottom trap with $|B_{min}| = 0$ adiabatically into a non-zero trap-bottom trap with $|B_{min}| = 0$

15. Towards: Cold atoms in a superconducting micro-trap

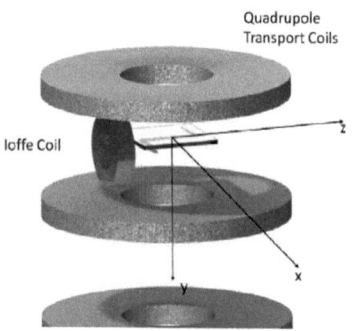

Figure 15.1: The reloading scheme is illustrated, showing three of the vertical transport coils, where the uppest generate the final quadrupole trap. The compact QUIC-coil in radial direction in addition is used to maintain a quadrupole-Ioe configuration trap which is the starting point for reloading into the micro trap. The chip-surface is illustrated transparent with a Z-shaped wire on top, mounted face-down. Even if homogeneous bias-fields are necessary to maintain a Ioe-like trap in the Nb-Z wire, the coils are not shown in the figure.

The open question addresses the following consideration: Which trap bottom must be applied in the QUIC-trap before reloading, that an possible opening of the trap does not cause the atoms to be lost during reloading?

For the subsequently presented reloading calculation an initial trap position $370 \mu m$ below the chip-surface is assumed, which leads together with the homogeneous bias-fields and a Z-current of $I_Z = 1A$ in a $50 \mu m$ wide wire to suitable trapping parameter in the beginning of the reloading.

The weakening of the trapping field caused by the *Meissner-eect* in superconductors should only come into play if the atoms are held closer to the surface than the reloading position of $370 \mu m$, in the order of about twice its width[2]. Therefore the modified trapping potentials due to the *Meissner-eect* can be neglected for the reloading calculations.

15.1.1. The initial QUIC-trap

Starting from the QUIC-trap, based on the geometries in section (13.4), the trap bottom is assumed to be $B_0 = 1G$. This trap bottom origins from the dierence of the field components in z-direction as described in section (5.3.3). The stability strongly depends on the QUIC-current, as mentioned in the design considerations (12.1) and as shown in the measurement section (14.4). To match the reloading positions, the chip trap should be assumed to be o-centered by Δy corresponding to the chip-trapping position, and in z-direction by $\Delta z \approx 6.7mm$ as the QUIC-trap is shifted from the vertical transport axis towards the QUIC-coil.

15.1.2. The final z-trap

Before considering the superconducting Z-trap acting as a simple normal conducting wire-trap (the trapping distance of $370 \mu m$ even satisfies this assumption), the critical current in the Z-wire should be considered. Depending on both, the temperature and the magnetic field,

---
[2]Oral communication with *A. Emmert* from the *Haroche group*, also refer to [103, 104, 105]

15.1. Re-loading atoms in a superconducting micro-trap

the critical current density decreases as the field strength is raised, or if the temperature approaches T_c, the critical temperature.

The critical current for a NbTi Z-wire of 50×2 μm^2 at $T = 4.2K$ is $I_C = 1A$ at a magnetic field of $B = 10000G$ [182], which is therefore far beyond the demands for the setup. Care has to be taken for Nb, which yields for the given wire structure in $I_C \approx 0.2 - 0.4A$ for $B = 1000G$, and $I_C \approx 0.3 - 0.6A$ for $B = 100G$, depending on the fabrication technique [203].

Since the current in the Z-structure will be limited it is set to $I_Z = 1A$. Assuming a length $l = 2mm$ of the central lead of the Z-structure and a bias field in x-direction of 5G, the trapping distance is set to $370\mu m$ below the chip. Together with an Ioe-field in z-direction of 0.35G, trap frequencies of $\approx 2\pi \times 159Hz$, $\approx 2\pi \times 153Hz$ and $\approx 2\pi \times 19Hz$ are obtained at this distance.

The gradient in y-direction is therefore strong enough to hold the atoms against gravity (15 G/cm). This trap therefore provides a harmonic trapping potential comparable to the QUIC-trap mentioned above. In addition the chip trap can be both, further compressed and brought closer to the chip-surface by lowering the trap bottom, increasing the chip current, or increasing the homogeneous bias field in x-direction.

15.1.3. Reloading to a micro-trap

The principle of the reloading-process follows a continuous scheme, turning on the micro-trap while the atoms are still trapped in the QUIC-trap and subsequently switching o the latter. As already mentioned the trap depth will be lowered during the reloading process. This is quite bad as the atoms must be cold enough, that they do not get lost. This opening origins from the field components in x- and y-direction and is not seriously eected by the z-component.

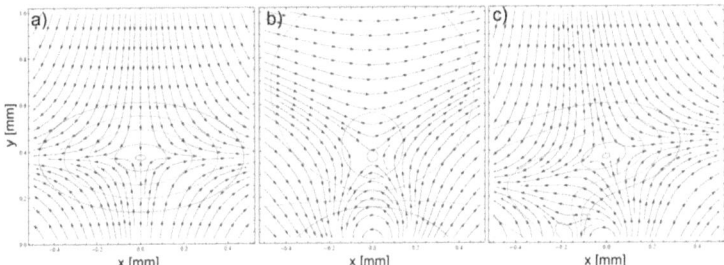

Figure 15.2.: Trap opening during reloading. Inset a) shows the magnetic field lines of the QUIC-trap or even a quadrupole trap maintained with the vertical anti-Helmholtz configured coils. Inset b) shows the magnetic field lines from the chip-based Ioe-trap, tilted by $\pi/4$ in respect to the quadrupole field from inset a). Both fields overlaid in inset c) show that the trap opens partially some when during reloading as both traps are maintained during the hybrid-trap phase which is demanding for a continuous, adiabatic reloading.

15. Towards: Cold atoms in a superconducting micro-trap

Partial trap opening during the reloading process

The main dierence of the Z- and the macroscopic coil-trap is that the quadrupole fields are rotated by an angle of $45°$ with respect to each other, if the usual bias field in x-direction is applied for the Z-trap. This is illustrated in Fig.(15.2) which shows the magnetic field in the x-y plane. Inset a) shows the quadrupole field generated by the quadrupole coils in the z=0 plane, with $I_{QUIC} = 0$. Inset b), shows the Z-trap with a negative Ioe-field so that it compensates B_z to zero. For negative x and small y values, the fields in inset a) and inset b) show approximately in opposite directions.
Both fields are superimposed in inset c). Therefore it can can clearly be seen that once in the reloading process the trapping potential is weakened as it is shown in the lower left corner of inset c). An additional field in z-direction generated by the Ioe-coil or the homogeneous Ioe-field would not influence this behavior meaningfully.

Tilting quadrupole fields

If the opening of the trap during the reloading process should be reduced, this can be done with higher currents especially in the Z-wire, however this current is restricted by a maximum current smaller than critical current I_c.
The alternative would be to rotate the quadrupole fields, to even reduce the angle between the fields. Therefore both fields can be rotated. Either the quadrupole coils are rotated around the z-axis (changing the complete transport geometry) or the direction of the Bias-field is rotated around the z-axis in order to rotate the quadrupole field of the trapping Z. Both alternatives faces drawbacks:

A) Rotating the transport coils by $45°$, leads to a closed trap with a trap hight of about $4G$ throughout the reloading process. Nevertheless the downside of such a setup is a more complicated transport into this trap and a worsened optical access.

B) The second possibility of rotating the bias field does not aect the optical access. However, in this scheme the final Z-trap is weakened. This is, because the trap center is together with the bias field rotated around the z-axis. The new location of the trap center has two disadvantages. First, it is closer to the chip surface, which should be avoided at least during the reloading process. Second, the bias field in y-direction is counteracted by the lead in x-direction of the Z-wire. For a rotation of $45°$, the trap hight is lowered by $\approx 1G$ compared to the above mentioned alternative A) with the bias-field in x-direction, but with a trap minimum even closer to the chip-surface.

The optimal scheme

Taking into account a partial opening trap during the reloading process with the original described geometric configuration, neglecting the more or less complicated alternatives A) and B), a further improvement of the trap depth can be achieved.
If the trap frequencies of both traps the initial macroscopic QUIC-trap and the final Z-trap are at least in the transversal directions the same, they can be kept constant during the reloading process ramping one trap down and the other at the same time up.

15.1. Re-loading atoms in a superconducting micro-trap

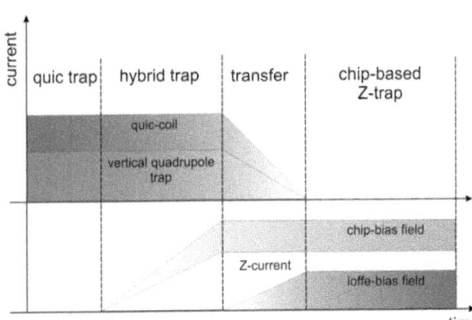

Figure 15.3: Current scheme for the reloading from a macroscopic quadrupole-ioe-configured QUIC-trap towards a chip-based microscopic Z-trap. The quadrupole components of the chip-trap are turend on to establish a hybrid trap at the matching position. Afterwards the ioe components are ramped up/down with the simultaneous switch o of the vertical transport coils to maintain a pure adiabatic reloading into a microscopic chip-based trap, even at a far-away trapping position.

For this scheme, Fig.(15.3) illustrates the time dependent currents in the coils. In addition this leads to an increase in the minimum trap depth during reloading process. The QUIC-trap is still maintained constant while the Z-structure and the bias-field are ramped up. For a lead length $l = 2mm$ of the chip-Z, the trap bottom automatically stays constant within certain limits if the initial and the final trap bottom are equal[3].

In a second step, the QUIC-trap is ramped down and at the same time the Ioe-field is ramped to its final value. The drawback of this scheme is, that the trap is slightly tightened and then released again during the reloading process.

The opening of the trap during the reloading process is shown in Fig.(15.4), inset a)-l). Inset b)-i) correspond to the hybrid-trap phase, where the chip-trap is established ramping up the Z-current and the homogeneous bias field. In contrast, inset j)-l) are related to the transfer phase where the Ioe-bias is ramped up, finally ending with the pure chip-trap.

The contour-plots are based on a 3D calculation and are cuts along the Ioe-direction (z-axis) through the longitudinal trap minimum. The calculation is done for a Z-length of $l_Z = 2mm$ at a distance of $370\mu m$ from the chip-surface which is the reloading position. The coils follow a setup as described in section 13.5, and the following magnetic field parameter with the coils built up by finite wires with the given number of windings, and the homogeneous fields assumed analytically. The chip wire is assumed to have the dimension of $d = 50\mu m$ and $h = 2\mu m$, which seems accurate enough at far-away reloading positions.

- $I_{quad} = 0.7A$, $N_{quad} = 3000$
- $I_{QUIC} = 0.832A$, $N_{QUIC} = 1800$
- $B_{Ioffe} = 0.35G$
- $B_{bias} = 5G$
- $I_Z = 1A$

---

[3] For small lead lengths the field in z-direction has to be compensated by the Ioe field, if the trap bottom is wanted to stay constant

15. Towards: Cold atoms in a superconducting micro-trap

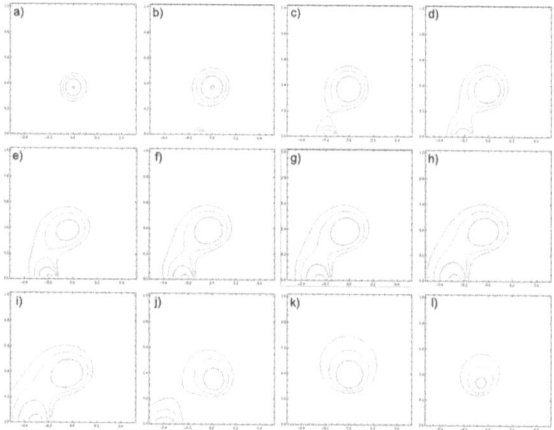

Figure 15.4.: Contour plots during the reloading process, starting from the QUIC-trap, inset a) towards a normal conducting Z-trap at a position $370\mu m$ away from the chip-surface. Inset b)-i) correspond to the hybrid-trap phase, whereas inset j)-l) are related to the transfer phase where the Ioe-bias is ramped up.

15.2. Trapping atoms in a superconducting chip-based micro-trap

From the final reloading position on, the atoms are trapped in a pure chip-based Z-trap. Possible materials for the chip-mounting and the chip-substrate have to follow the design-considerations as they are given in chapter (12).

Atom chip mounting and chip layout

An optimal chip-mounting would therefore provide a high thermal conductivity, and non-zero electrical resistance, to insulate the wires and structures from each other without the need for insulation. Hence, sapphire or even quartz[4] are such candidates. Fig.(15.5) shows a possible chip-mounting as it would fit into the setup, with the corresponding preliminary chip-layout. Inset a) shows the mounting, with an overall length of 58.5mm, a head-diameter of 36mm, and a bottom diameter of $D_B = 46mm$. The head is truncated, to let space for the QUIC- and super-Ioe-coil. As sapphire is a very brittle material, drilling tests, inset c), were done, to see if the desired holes can be drilled as shown in inset a).

---

[4] For further detail on the material properties see the Appendix. Recent calculations have shown, that a desired quartz-crystal substrate would just need $2 - 4\times$ longer than Cu to be cooled down

15.2. Trapping atoms in a superconducting chip-based micro-trap

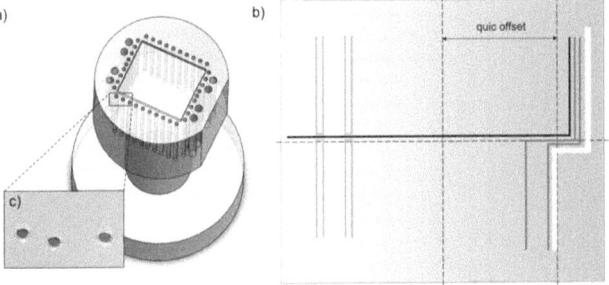

Figure 15.5.: A sapphire mounting inset a), and a chip design are illustrated. The substrate keeps two Z-shaped trapping wires (with central lead-length 2mm, and 3mm), an RF wire next to them, and a guiding wire from right to left with 4 U-shaped end-caps for the wire trap. The wire-Zs are o-centered from the vertical axis, as the trapping position in the macroscopic Ioe-like trap is shifted towards the QUIC-coil. Inset c) shows a drilling test at a sapphire substrate which indicates that drilling must even be improved to avoid shivering of the sapphire.

Chip based Z-trap

This configuration provides a standard chip trap capable for cooling and compressing the atoms, even to reach the critical phase space density for Bose-Einstein condensation [188]. Even the position of the trap is at a far-away distance of $370 \mu m$ from the chip-surface after successful reloading, the atoms can be brought closer to the surface by increasing the chip-bias field or even the vertical bias field.

Accurate calculations therefore must follow [103, 104, 105], and section (5.3.4), to consider the changed trapping potentials due to the current density, influenced by the *Meissner-e ect* . A possible chip-design is shown in inset b) with two dierent wire traps, based on a Z-structure with an lead length of 2mm, respectively 3mm in addition with a broad wire (yellow) for shining in the RF-cooling power. A wire-guide therefore would allow to transport the atoms to a final position fixed with 4 U-shaped wires shown on the left end of the chip for further manipulation and coupling to a microwave resonator.

Part III.
Conclusion and Outlook

16 Conclusion

Electron-gun

To conclude, I presented an electron gun, and a specific designed electrostatic lens-system which is capable of focusing the electron beam to the sub-mm regime. The e-gun can be operated both with a W- or a PtIr emitter, in the cold field emission regime. It provides emission currents up to $100\mu A$ and 6kV beam energy, and withstands at least 2000 cycles with $I_e = 10\mu A$, each cycle lasting for $t \approx 2s$. A subsequent deflection system provides full flexibility to scan the electron beam along the target-surface.
Even if the electron beam crossed the MOT-region were magnetic quadrupole fields occur, the beam was not essentially perturbed by the fields.

Fur future applications of the e-gun and the electrostatic lens system, a redesign of the latter would be advantageous to enable even a focused spot-size in the order of 1-50 μm instead of being limited at $r_{spot} =\approx 200 - 300 \mu m$.

E-beam driven atom source

The performed measurements clearly showed that the electron beam is responsible for the loading of the MOT. Compared to classical, resistive heated dispensers, the presented alkali metal atom-source provides a loading rate which is more than a factor of 1000 bigger, when scaled by heating power. In a loading time of $t = 1s$ the ebeam-driven MOT is therefore capable of trapping $\approx 4 \times 10^6$ atoms.
Even with a beam power of $200\mu W$ and a completely divergent e-beam, a loading rate of roughly 7×10^5 atoms/sec could be achieved, leading during the 1.5s lasting loading phase to 1.110^6 atoms trapped in the MOT. This is a strong evidence that the underlying process for desorbing the atoms can not be electron beam evaporation of alkali metals. The corresponding scaling laws for the experimental parameter were therefore derived in the conceptual section where this process is under the looking-glass. It rather indicates that ESD is the fundamental process, indicated by the linear increase of atoms trapped, with target current. Another indication is the pressure rise in the vacuum-chamber measured with the quadrupole mass spectrometer. The rise in hydrogen-, nitrogen-, and carbon-dioxide partial pressures are known from high energy particle physics, where electron-stimulated desorption

16. Conclusion

together with PSD are responsible for bad pressure in the system.

This indeed was the limiting factor for the electron-beam loaded MOT, introduced in this thesis. As a non-negligible fraction of the emitted electron current got lost in the lens system this lead to an remarkable rise in the pressure. This pressure rise strongly limited higher loading times as the background pressure induced trap losses in the MOT and prevents the system to be scaled to higher emission currents. In addition a reliable method of target preparation with simultaneously the possibility to determine the target thickness would have been of advantage.

Even if the desired number of trapped atoms was not reached, this experiment exhibited a future application to setups which demand less atom numbers, even in cryogenic systems. With its low power consumption in the region of $10^{-3}W$ and less, it can even be implemented in cryogenic dilution systems which reaches milli-Kelvin temperatures. Recent experiments also exhibit the possibility to use an electron beam to prepare ions based on a Rb-MOT [204]. With an electron-beam driven atom-source for alkali-metals loading a MOT, as presented in this work, even ions in cryogenic systems could easily be prepared.

Transport of cold atoms inside a 4K environment

Experimental experienced, the magnetic transport is extremely robust, both in variations of the transport currents and the repeating opening and closing of the vertical cryogenic transport section.

In addition to the well known horizontal transport, a novel vertical transport scheme was developed, even with superconducting coils in the cryogenic section. The cryogenic setup provides flexible trapping geometries and is capable to implement a superconducting atom chip at $T \approx 5K$. Optical access allows to take pictures of the trapped atom-cloud. Hence, the setup is capable to transport more than 7×10^7 cold atoms in a cryogenic environment, finally trapped in a macroscopic superconducting quadrupole-trap with lifetimes exceeding 230s. Furthermore the atoms were trapped in a macroscopic sc-Ioe-quadrupole like trap, which allows for further cooling and enhanced trapping densities. This will act as a suitable starting position to reload atoms to a sc-chip based micro-trap for further manipulation.

Beside the challenging fabrication of the highly inductive superconducting coils, their intrinsic properties, namely their inductance and capacitance, and a non-simultaneous quenching of the trapping currents, obviously caused oscillating magnetic fields after switching o the trapping fields. This leads to time-dependent Zeeman-shifts of the hyperfine sub-levels mF and makes imaging, as well as extraction of information from TOF pictures challenging. Nevertheless this could be prevented introducing a serial circuit for the QUIC-trap, which would increase trap stability and possibly avoid oscillating magnetic fields after switch-o.

In addition, this setup is fully flexible and allows even for reloading the atoms onto a sc-surface trap, or even bringing the atoms close to the surface for further investigations.

17 Outlook

Beside evaporative cooling [42] and Bose-Einstein condensation [2, 1], in the superconducting chip-based micro-trap or even the macroscopic QUIC-trap, the presented flexible setup allows for the investigation of surface eects on cold quantum gases [11] and investigation of cold atoms close to superconducting surfaces [24, 205, 25] as referenced.

In addition this setup opens an avenue for coupling an ensemble of cold atoms to superconducting microwave resonators at $T = 4K$ [206], and testing this scheme for establishing cold atoms inside a cryogenic environment even for systems much below $T = 4K$. Those proposals, aiming for the realization of quantum memories [35, 207, 208, 31] with a trapped ensemble close to a superconducting surface, opens the road for novel experiments, which even demands temperatures in the mK-regime[1], and addresses the demand for a e cient way of transporting cold atoms into a cryogenic system.

In parallel to the presented work, the above mentioned superconducting resonators are currently developed at the *Vienna Center for Quantum Science and Technology, Atominstitut - Vienna/Austria*. Fig.(17.2), inset a) therefore shows a closeup of a superconducting resonator taken with an optical microscope, and inset c) exhibits a resonator in a sample-mount prepared for the 15mK-dilution cryostat. A principle sketch, of an trapped atomic ensemble close to a resonator is illustrated in inset b) with the red cloud close to a coplanar MW resonator. The following describes how magnetic coupling to a planer resonator-device is achieved.

From cavity QED to circuit-CQED

Considering a single two-level system, the coupling to a single-mode cavity field is described by the *Jaynes-Cummings model* [209]. For a standing wave, in between two mirrors, the system is illustrated in Fig.(17.1), inset a), and beside the coupling-constant g, the loss from the cavity depends on the quality-factor Q, written as $= \,_{cav}/Q_{tot}$. The emission rate is therefore written as .

For coupling to a resonator in the MW-regime, the mode volume is in principle huge compared to the dimensions of the single two-level system, nevertheless a case as depicted in

---

[1] Such experiments could hardly be performed at the setup described in this thesis, nevertheless such a transport scheme could be installed quite easily in a dilution-fridge operated at mK-temperatures

17. Outlook

Fig.(17.1), inset b), can be considered [210] which decreases the mode-volume, and increases the coupling dramatically. Considering electric coupling, simple calculations [211, 212] can be performed which exhibits that the maximum vacuum-Rabi-frequency, the coupling-strength of the transition to the cavity, in principle is just limited by fundamental constants.

As the electric field amplitude of the vacuum fluctuations E_0 is that due to half a photon and the energy of the photon is stored equally in the electric and the magnetic field respectively, the spatial integral of the electric field energy can be written as

$$\frac{\hbar \omega_{cav}}{4} = \frac{\epsilon_0}{2} \int E_0^2 dV \qquad (17.1)$$

with the cavity frequency ω_{cav}, the vacuum permittivity and the mode-volume V. Considering electric coupling, the coupling strength can be written, expressing the dipole moment in terms of the dimension L of the single two-level system which reads as

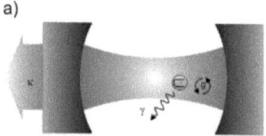

$$g = \frac{d \cdot E_0}{\hbar} \quad \text{with:} \quad d = e \cdot L \qquad (17.2)$$

With Eq.(17.2) as an expression of E_0^2, Eq.(17.1), can be rewritten as

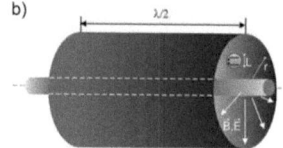

$$\frac{\hbar \omega_{cav}}{4} = \frac{\epsilon_0}{2} \frac{g^2 \hbar^2}{d^2} V \qquad (17.3)$$

For a cavity with a transversal dimension r, the volume can be written as $V = r^2 \cdot \lambda/2$ if the length of the cavity, and hence the resonator-length is $\lambda/2$ as depicted in Fig.(17.1), inset b). Therefore the ratio g/ω_{cav} between coupling strength and cavity frequency can be expressed using Eq.(17.3) and reads as

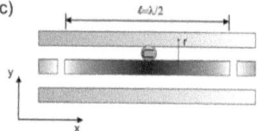

Figure 17.1.: From cavity-QED to circuit CQED

$$\frac{g^2}{\omega_{cav}^2} = \frac{e^2 L^2}{2 \cdot 2 r^2 \epsilon_0 \hbar c} \qquad (17.4)$$

and hence using the fine-structure constant α as

$$\frac{g}{\omega_{cav}} = \sqrt{\frac{\alpha}{2}} \left(\frac{L}{r}\right) \qquad (17.5)$$

if the cavity wavelength is written as $\lambda = 2c/\omega_{cav}$ and the cavity volume therefore is written as $V = r^2 \cdot 2 \frac{c}{\omega_{cav}}$.

166

For a planar geometry such as depicted in Fig.(17.1), inset c), a high coupling constant g can be achieved if the cavity is restricted in the transversal direction.

Magnetic coupling to superconducting coplanar MW-resonator

A superconducting atom-chip therefore provides a platform which is well suited to implement such a planar superconducting coplanar MW-resonator [14] as it would be demanded for experiments following [31, 33], coupling N ultra-cold atoms to a superconducting planar microwave resonator.

The photons bouncing back and forth in the resonator corresponds to a standing electromagnetic wave in the GHz-regime which provides locally strong magnetic fields, as even shown for the electric field component in the section above, but also valid without restrictions for the magnetic energy in a restricted mode-volume.

Based on the principle spin-oscillator Hamiltonian of the system $H = H_{atom} + H_{cav} + H_{int}$, the interaction term can be written as

$$H_{int} = \mu \cdot B = \frac{\mu_B}{\hbar}\left(g_S S - \frac{\mu_N}{\mu_B} g_I I\right) \cdot B \tag{17.6}$$

were the magnetic momentum interacts with the cavity magnetic field B. For the corresponding magnetic dipole transition, the interaction μ of the cavity with the the trappable hyperfine states of ^{87}Rb, namely the $5^2 S_{1/2} |F=1\rangle \to 5^2 S_{1/2} |F=2\rangle$ transition at $\omega = 6.83 GHz$ as shown in Fig.(17.2), inset d), can be calculated.

Collective and strong coupling

Nevertheless for magnetic interactions, the coupling of a single rubidium atom is quite weak, with g_0 in the order of 40Hz [31] for a distance as close as several micrometers above the central conductor, and much smaller than the cavity losses $\kappa = \omega_{cav}/Q \approx 2\pi \times 7kHz$ and the decay rate $\gamma \approx 2\pi \times 0.3Hz$ [2].

The week coupling can obviously be overcome by coupling an ensemble of atoms to the cavity mode, as the single atom coupling strength g_0 for coupling an ensemble of atoms enhances, written as $g_{eff} = \sqrt{\sum_i g_i^2}$ following [213, 214, 215]. Described by the *Tavis-Cummings model* [216], the interaction Hamiltonian can therefore be rewritten

$$H = \hbar \omega a^\dagger a + \hbar \omega_a \tilde{\sigma}^{+-} + \hbar g_{eff}\left(\tilde{\sigma}^{+} a + a^\dagger \tilde{\sigma}^{-}\right) \tag{17.7}$$

with the single coupling strength $g_i = [b_i(\bar{x},\bar{y}) \cdot \mu_i]/(\sqrt{2}\hbar)$ for a mode-function $b_i(\bar{x},\bar{y})$ of the magnetic field strength and the transition element μ_i for the magnetic dipole transition as described in Eq.(17.6). The first term in Eq.(17.7) therefore describes the cavity, with the photon annihilation and creation operators a, $a^\dagger$, while in the second and third term the normalized collective excitation operators $\tilde{\sigma}^{\pm} = \frac{1}{g_{eff}} \sum_i g_i \sigma_i^{\pm}$ which describes the creation or annihilation of a single excitation in the atomic ensemble, enters. Therefore, $g_{eff} = \sqrt{N} \cdot g_0$

---

[2]NOTICE: For $\kappa \approx 2\pi \cdot 7kHz$, a quality factor of 10^6 is assumed, where the decay rate $\gamma \approx 2\pi \cdot 0.3Hz$ is limited by the trapping lifetime

17. Outlook

holds if the single-atom coupling constants g_i are equal. The single excitation in an atomic ensemble therefore is described by a normalized *Dicke-state* which looks like

$$\frac{1}{\sqrt{N}} \sum_{i=1}^{N} |\downarrow \ldots \uparrow_i \ldots \downarrow\rangle \qquad (17.8)$$

Finally this is a major reason why a collective excitation of a single photon in an atomic ensemble exhibits a quantum memory of extraordinary robustness, as the loss of a single atom, negligible perturbs and destroys the original *Dicke-state*.

Usually the cooperativity-parameter $C = \frac{g^2}{\kappa\gamma}$ describes collective phenomena, nevertheless it is a meaningless parameter since is small anyway. Magnetic strong coupling of the ensemble to the cavity-mode can therefore be achieved, if the eective coupling strength g_{eff} overcomes the losses of the resonator described by $g_{eff} \gg \kappa$ and $g_{eff} \gg \gamma$ anyway, which could be achieved as an ensemble of 10^6 atoms would result in $g_0 \to g_{eff} = \sqrt{N} \, g_0 \approx 40 kHz$.

Restrictions to collective coupling

As just the low field seeking states of ^{87}Rb are trappable, a two photon transition has to be taken into account [39]. Introducing a virtual level between $|F = 2, m_F = 0\rangle$ and $|F = 2, m_F = 1\rangle$, as depicted in Fig.(17.2), inset d), the eciency of the coupling and the dephasing, would strongly depend on the absolute stability of the transitions and the introduced detuning $\Delta = \omega - \omega_a$.

The reason is twofold, as both external magnetic field such as trapping fields would cause a shift in the Zeeman-sublevels for $m_F = -1$ and $m_F = +1$ and therefore in ω_a, and a shift in the cavity-frequency ω_c would additional introduce a change in detuning Δ. Notice that the collective coupling g_{eff} as described above does not yet consider the introduced detuning Δ as demanded for a two-photon transition.

The first constraint, regards to the magnetic trapping potentials which vary with trapping position if the atoms are not cold enough[3], while in addition the mode-function $b(x, y)$ of the magnetic field strength would change spatially, which would make the assumption $b(\bar{x}, \bar{y}) = b(x, y)$ invalid.

The second constraint, is the main-reason why the complete transport setup is built without the use of an optical transport scheme, as investigations[4] exhibited a resonator shift of at least $10 kHz/\mu W$ light power and an additional loss in quality-factor of the resonator, and hence increase in cavity loss rate κ, which does not recover with time, except heating above T_C.

In addition Bose-Einstein Condensation is not an essential condition, as the enhancement by $\sqrt{N}$ does not depend on the temperature. Nevertheless a compact and dense atomic cloud is demanded which limits spatial inhomogeneous *Zeeman-shifts* below a critical value. Both, the light induced frequency shift, and the spatial inhomogeneous coupling would decrease coupling eciency.

As mentioned above, the Q-factor is crucial for achieving strong coupling as it keeps the cavity

---

[3] As the chip-trap provides a harmonic potential, atoms at the outer end of the trap are located at higher magnetic fields, than in the center

[4] These investigations are done for niobium MW-resonators with Q-factors up to 900.000

losses small. Studies which investigates perpendicular magnetic fields on superconducting Nb-films and resonators, as well as occurring vortices, could therefore follow work done in [217].

Super-radiant decay of atoms in a MW-cavity

Experimental prospects following [206] even at temperatures of 4K, are based on an experimental setup as described in this thesis, which demands in addition a superconducting microwave resonator to be established. This system would it make possible, to buiuld a narrow-band strip-line micro-maser, locked to the atomic hyperfine transition of the trapped atoms, following [206]. In addition collective phenomena such as super-radiant decay could be studied. An inverted ensemble which is initially prepared in

$$\frac{1}{\sqrt{N}} \sum_{i=1}^{N} |\uparrow ... \uparrow_i ... \uparrow\rangle \qquad (17.9)$$

leads to a radiative burst, which could be easily observed in the over-damped regime ($\gg g_{eff}$), as all photons would leave the cavity and the atomic ensemble decays into the collective ground state

$$\frac{1}{\sqrt{N}} \sum_{i=1}^{N} |\downarrow ... \downarrow_i ... \downarrow\rangle \qquad (17.10)$$

The collective atomic sample would therefore emit the photons on a time scale $t \propto 1/(N\Gamma_0)$ with the free-space spontaneous emission rate Γ_0 and the *Purcell-factor*, so that the cavity stays empty at all times[5].

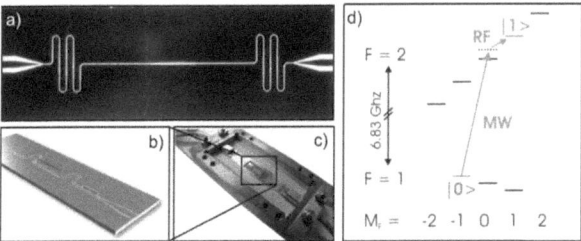

Figure 17.2.: Inset a) depicts an overview image of a micro-fabricated niobium resonator with a bent wire structure. Inset b) illustrates an ensemble of cold atoms in a typical cigar-shaped micro-trap close to the central-wire of the resonator with a superconducting solid-state qubit located at the right end of the resonator. Inset c) shows a typical setup how those resonators are mounted in a sample-holder for testing at $T = 30mK$. The right inset d) shows the clock-state transitions for magnetically coupling the atomic qubit to the resonator transition. To allow for an excitation from the ground-state $|g\rangle$ to an excited state $|e\rangle$ of the atomic ensemble with a resonator photon, a two photon transition has to be applied. The detuning of the virtual level to the atomic transition would in addition decrease the coupling-e ciency.

---

[5]For further details on superradiance, the reader is redirected to literature on CQED, such as [211]

Part IV.

Addenda

18 Sincerce thanks are given to...

- Jörg Schmiedmayer, who gave me the opportunity to do my research and thesis within the Quantum Optics Group and who let me join a row of challenging experiments. I always enjoyed copious discussions and bringing forward new ideas.

- Stephan *Mr. Vacuum* Schneider, and Hannes Majer, the mighty wizard of electronics, which both additionally supervised my thesis and kept my back free. Thanks for providing me with a bunch of helpful gadgets and all the technical support. In addition I want to mention Martin Trinker and Jose Verdu, another two more PostDocs, which partially supervised my work.

- Robert Amsüss, who joined me quite a while for a challenging project, and eventually supported me to finalized it. Robert, it was quite a fun, and I will not forget the hot tempered, but nevertheless fruitful discussions.

- Christoph Hufnagel-San, the man with mighty hands which showed outstanding endurance in the lab, participating a lot in getting the beast working. Unforgettable how you widened my horizon for exceptional musical genres.

- The chocolate fairy, also known as Christian Koller. Thanks for having always an open ear, desk lay and ice chest.

- The cumulative QUIC-team: You inspired my work by believing in an outstanding and challenging experiment. Christian, you did manage to do your master thesis really well, therefore thanks and admission for your engagement. Thanks to Fritz Diorico and Stefan Minniberger. One day you will reap the fruits of the past. You managed quite well to learn a lot during past month and it is for sure, you will tame the beast! You really teamed up with each other, and I wish you all the best. In addition I want to mention Nils Lippok, who set up parts of the magnetic transport during his diploma-thesis.

- My o ce mates, Stephanie, Thomas, Christoph, Wolfi, Matthias and others more. It was a great pleasure with you. Even thanks for amusement and all the golden hours. Especially thanks to Wolfgang for the straight forwarded help at the ADwin.

18. Sincerce thanks are given to...

- H. Hartmann, A. Linzer and all the other guys from the mechanical workshop. You always did your best, and supplied me rapidly with prototypes. Thanks for your support and some brilliant ideas you came up with.

- J. Summhammer, how let me exploit his old machine, and to T. Juman for a tutorial in etching W-tips.

- Michel Brune for being my second examiner, and to Stephan Schneider, Hannes Majer, and Robert Amsüss for proof-reading this thesis.

- My colleagues, and all other group members, for any kind of support and all the coee-corner adventures.

- My dear friends, for simply being my friends and all the support, especially Johannes and Tom for the times you had to get on without me.

- My family, and beyond that especially Inge, Hans, Georg, Johanna, Cleide, Christine, Felix, Katharina, Lukas and Christoph. Thanks for your support, during the past years, my studies and even before. Thanks for getting along with me, or even without me.

- Elisabeth. You were right - Mphcev. Thank you for everything, the paella, and infinitely more.

List of Figures

5.1.	Probing techniques with electron beams, exhibiting electron-bulk interactions	14
5.2.	Definition of the *Kanaya-Okayama* penetration range for electrons in bulk-material	16
5.3.	Fractional backscattering of electrons in a bulk	17
5.4.	Different field configurations for the interaction of electrons with electrostatic and magneto-static fields	19
5.5.	Setup of a QUIC-coil	25
5.6.	Simulation of a quadrupole-Ioffe-configuration (QUIC-) trap configuration	30
5.7.	Lifetime in dependence of different losses	31
6.1.	Vapor pressure for different elements	37
6.2.	Simulation of the penetration depth of electrons in a rubidium-layer	38
6.3.	Comparison of the calculated *Kanaya-Okyama* range and the simulation of the penetration depth	39
6.4.	Target temperature with spot size	41
6.5.	Heating rate on target	42
6.6.	Estimation of the atom flux in the MOT region after heating the target with an electron beam	44
6.7.	Simulation of the effective spot-size	45
6.8.	Fraction of trappable atoms due to Maxwell-Boltzmann velocity distribution	47
6.9.	Trappable atoms: Study on the dependence of beam current and spot size	49
6.10.	Trappable atoms: Study on the dependence of target temperature and spot size	49
7.1.	Conceptual setup of an electron beam loaded MOT	51
7.2.	Field emission current for different work-functions	53
7.3.	Images of different emission tips	54
7.4.	Principle setup of the electron source with an archetype lens system	56
7.5.	Illustration of the e-beam crossing the MOT region, possibly leading to ionization events in the MOT	57
7.6.	Simulation of different electron trajectories passing a magnetic gradient field	58

List of Figures

7.7. Perturbation simulation of electrons crossing a quadrupole field region 59

8.1. Laser setup for loading a MOT as used in the e-beam experiment 64
8.2. Emission characteristic of an etched W-tip 65
8.3. Illustrated cut through the electron-gun 66
8.4. Assembled electron-gun 67
8.5. Dierent e-gun targets 68
8.6. Experimental setup of the electron-beam MOT experiment 70

9.1. Focusing behavior of the electron-gun without magnetic fields 74
9.2. Transfer e ciency between emitted current and target current in the presence of the MOT-quadrupole field 75
9.3. Atoms in the MOT before and after switching on the electron-gun 76
9.4. Experimental cycle for measuring the loading rate and atom-number in the electron-beam driven MOT 77
9.5. Comparison of the loading rate and desorption yield of atoms trapped in the MOT, for the electron gun switched ON/OFF 78
9.6. Unfocused desorption, and trapped atom-number and target current respectively, in dependence on the focusing voltage 79
9.7. Comparison of the atom-number and the desorption yield for dierent target temperatures 80
9.8. Pressure increase measured with QMA 82

10.1. Dierent setups for a magnetic transport line 86
10.2. Horizontal transport scheme for constant trap gradients 90
10.3. Vertical transport scheme for constant trap gradients 91
10.4. Coil setup for the magnetic transport 92
10.5. Aspect ratio of the magnetic transport at the beginning and the end of the scheme 94
10.6. Transport currents for the horizontal transport 95
10.7. Vertical transport currents 96

11.1. D2-line of ^{87}Rb with the cooling, repumping, imaging and optical pumping transitions 101
11.2. Locking-scheme of the frequency oset (FO) lock 102
11.3. Locking scheme of the cooler-laser 102
11.4. Optical setup of the repumper-laser *Rudi* 104
11.5. Laserpark of the experiment 105
11.6. Vacuum chamber setup for the lower magnetic trap 107
11.7. Setup of the horizontal transport line 108
11.8. Setup of the MOT-optics 109
11.9. Temperature sensors for the magnetic transport line 110
11.10 Experimental setup of the lower transport chamber and the horizontal transport line 111

12.1. Proposed cooling power of the DE210S on the second stage 116
12.2. Illustration of the implementation of a QUIC-coil 119
12.3. Optimization of the wire parameter 122

List of Figures

12.4. Influence of the thickness of the radiation shield on the cooling-performance . 126

13.1. The cold-finger and the vacuum-chamber 128
13.2. The thermal radiation shield 129
13.3. Wires thermally anchored and soldered at the first cooling-stage 131
13.4. Experimental 4K-plate with interior 132
13.5. Temperature sensors at 4K 133
13.6. Overview of the 4K environment 133
13.7. Superconducting coils and Cu-mountings 136
13.8. Experimental setup of the two chamber setup with the magnetic transport line 137
13.9. Overview of the assembled setup for transport of atoms into a cryogenic environment 138

14.1. TOF scan of atoms trapped in the superconducting quadrupole trap 141
14.2. Oscillating magnetic field after switching o the superconducting quadrupole trap 142
14.3. Comparison of absolute position for TOF imaging after sc-quadrupole trap . 143
14.4. Lifetime and heating rate in the lower magnetic trap 144
14.5. Transport e ciency for the horizontal section 145
14.6. Atoms in the sc-quadrupole trap depending on vertical transport parameter, I 146
14.7. Atoms in the sc-quadrupole trap depending on vertical transport parameter, II 147
14.8. Optimizing atom number in the cryogenic environment 148
14.9. Lifetime of atoms in the superconducting quadrupole trap 149
14.10 In-situ images exhibiting iso-B-field lines in the superconducting quadrupole trap 150
14.11 Comparsion of measured and calculated position shift of the superconducting quadrupole trap 151
14.12 Lifetime of the atoms in the superconducting QUIC-trap 152

15.1. Reloading scheme to load cold atoms into a superconducting micro-trap based on an atom chip 154
15.2. Partial opening of the trap during reloading 155
15.3. Current scheme for the reloading towards an atom chip 157
15.4. Contur plots during the reloading process 158
15.5. Mounting and chip design 159

17.1. From cavity-QED to circuit CQED 166
17.2. Superconducting Micro-Wave resonators 169

A.1. Compressor and cryo-cooler scheme 200
A.2. Heat load performance of the GMX20-B system 201
A.3. Base temperature at the 2nd-cooling stage during the experimental cycle . . 202
A.4. Window mountings at the thermal shield 203
A.5. Superconducting transport coils 204
A.6. Avalanche eect at the solder-contact due to heat dissipation into the sc-wire 205
A.7. Transport currents for testing the robustness of the vertical transport 208
A.8. Results of the robustness testing for the vertical transport coils 209
A.9. Measurement of the oscillating current and comparison with atom-number . . 213

List of Figures

A.10.Imaging setup 217

B.1. Electronics - switch-box for the magnetic transport 220
B.2. Scheme of the MOSFET-benches 221
B.3. Circuit diagram of the de-multiplexer board 222
B.4. Circuit diagram of the demux-control board 223
B.5. Circuit scheme of the superconducting coil switch 224

C.1. Experimental control setup 226
C.2. Flow chart for the experimental control 227

E.1. Relevant transitions of ^{87}Rb 233

G.1. Technical Construction Drawing: Push-coil winding scheme 237
G.2. Technical Construction Drawing E-gun: Overview of the electron-gun 238
G.3. Technical Construction Drawing E-gun: Field emission tip-holder and deflection plates 239
G.4. Technical Construction Drawing E-gun: Electrodes of the lens-system 240
G.5. Technical Construction Drawing E-gun: Various macor-insulators 241
G.6. Technical Construction Drawing E-gun: Overview of the inner life of the LN_2 target flange. 242
G.7. Technical Construction Drawing: Cryostat vacuum-chamber 243
G.8. Technical Construction Drawing: Upper thermal shield 244
G.9. Technical Construction Drawing: Lower thermal shield 245
G.10.Technical Construction Drawing: Overview of the thermal shielding at the 1st-stage 246
G.11.Technical Construction Drawing: 1st-stage window mounting (99.5 Al) 247
G.12.Technical Construction Drawing: 1st-stage window tightening steel-ring . . . 248
G.13.Technical Construction Drawing: Cu-base to sustain the 4K experimental setup 249
G.14.Technical Construction Drawing: Vertical Cu-rods for sustaining the coil-cage 250
G.15.Technical Construction Drawing: Transport-coil mounting 251
G.16.Technical Construction Drawing: Ioe-coil mounting 252

List of Tables

8.1. Maximum and typical operating potentials applied at the e-gun electrodes . . 67

10.1. Dimensions of the horizontal transport coils 92
10.2. Dimensions of the vertical transport coils 93

11.1. A liation of switch-boxes, coils and power supplies 111
11.2. Experimental cycle for preparing the atoms for transport 112

12.1. Influence of the Ioe-current in a QUIC-trap on the trap bottom and the transversal trap frequencies 118
12.2. Overview of the heat budget for the wiring 121

13.1. Wires installed for use in the cryogenic 4K setup 130
13.2. Dierent coil, and coil-mounting types for the sc-vertical-transport 134

14.1. Deviation from preset transport currents 140
14.2. Optimizing atom number in the cryogenic environment 148

A.1. Real coil dimensions of the superconducting transport coils 210
A.2. Trials to shape superconducting vertical transport currents 210

D.1. Thermal and electrical properties of metals 229
D.2. Thermal and electrical properties of Glues & Grease 230
D.3. Thermal properties for chip mountings 230
D.4. Thermal and electrical properties of used materials 231

E.1. Physical properties of ^{87}Rb, taken from [140]. 234
E.2. Optical properties of ^{87}Rb, and important deduced parameter for the D2-line $(5^2S_{1/2} \rightarrow 5^2P_{3/2})$, taken from [140]. 234

F.1. Physical Constants 235
F.2. Deduced Physical Constants 235

Bibliography

[1] E. A. Cornell and C. E. Wieman. Nobel lecture: Bose-einstein condensation in a dilute gas, the first 70 years and some recent experiments. *Rev. Mod. Phys.*, 74:875–893, 2002.

[2] W. Ketterle. Nobel lecture: When atoms behave as waves: Bose-einstein condensation and the atom laser. *Rev. Mod. Phys.*, 74:1131, 2002.

[3] S. N. Bose. Plancks gesetz und lichtquantenhypothese. *Zeitschrift für Physik*, 26:178, 1924.

[4] A. Einstein. Quantentheorie des einatomigen idealen gases. *Sitzungsber. Kgl. Preuss. Akad. Wiss*, 261:3, 1924.

[5] J. Denschlag, G. Umshaus, and Jörg Schmiedmayer. Probing a singular potential with cold atoms: A neutral atom and a charged wire. *Phys. Rev. Lett.*, 81:737, 1998.

[6] R. Folman, P. Krüger, J. Schmiedmayer, J. Denschlag, and C. Henkel. Microscopic atom optics: from wires to an atom chip. *Adv. At. Mol. Opt. Phys.*, 48:263, 2002.

[7] J. Fortagh and C. Zimmermann. Magnetic microtraps for ultracold atoms. *Rev. Mod. Phys.*, 79:1–55, 2007.

[8] R. Folman, P. Krüger, D. Cassettari, B. Hessmo, T. Maier, and J. Schmiedmayer. Controlling cold atoms using nanofabricated surfaces: Atom chips. *Phys. Rev. Lett.*, 84:4749, 2000.

[9] J. Reichel. Microchip traps and bose-einstein condensation. *Appl. Phys. B*, 74:469–48, 2002.

[10] M. Trinker, S. Groth, S. Haslinger, S. Manz, T. Betz, I. Bar-Joseph, T. Schumm, and J. Schmiedmayer. Multilayer atom chips for versatile atom micromanipulation. *Appl. Phys. Lett.*, 92:254102, 2008.

[11] S. Wildermuth, S. Hoerberth, I. Lesanovsky, I. Bar-Joseph, S. Groth, P. Krüger, and J. Schmiedmayer. Sensing electric and magnetic fields with bose-einstein condensates. *Appl. Phys. Lett.*, 88:264103, 2006.

Bibliography

[12] M. Wilzbach, D. Heine, S. Groth, X. Liu, T. Raub, B. Hessmo, and J. Schmiedmayer. Simple integrated single-atom detector. *Optics Letters*, 34:259–261, 2009.

[13] D. Heine, M. Wilzbach, T. Raub, B. Hessmo, and J. Schmiedmayer. Integrated atom detector: Single atoms and photon statistics. *Phys. Rev. A*, 79:021804(R, 2009.

[14] L. Frunzio, A. Wallra, D. Schuster, J. Majer, and R. Schoelkopf. Fabrication and characterization of superconducting circuit qed devices for quantum computation. *IEEE Transaction on applied superconductivity*, 15:860–863, 2005.

[15] A. Wallra, D. Schuster, A. Blais, L. Frunzio, R.-S. Huang, J. Majer, S. Kumar, S. Girvin, and R. Schoelkopf. Strong coupling of a single photon to a superconducting qubit using circuit quantum electrodynamics. *Nature*, 431:162, 2004.

[16] D. Schuster, A. Wallra, A. Blais, L. Frunzio, R.-S. Huang, J. Majer, S. M. Girvin, and R. J. Schoelkopf. Ac-stark shift and dephasing of a superconducting quibit strongly coupled to a cavity field. *Phys. Rev. Lett.*, 94:1236, 2005.

[17] P.K. Day, H. G. LeDuc, B. A. Mazin, A. Vayonakis, and J. Zmuidzinas. A broadband superconducting detector suitable for use in large arrays. *Nature*, 425:817, 2003.

[18] T. Nirrengarten, A. Qarry, C. Roux, A. Emmert, G. Nogues, M. Brune, J.-M. Raimond, and S. Haroche. Realization of a superconducting atom chip. *Phys. Rev. Lett*, 97:200405, 2006.

[19] T. Mukai, C. Hufnagel, A. Kasper, T. Meno, A. Tsukada, K. Semba, and F. Shimizu. Persistent supercurrent atom chip. *Phys. Rev. Lett*, 98:260407, 2007.

[20] J. Mozley, P. Hyafil, G. Nogues, J.-M. Raimond M. Brune, and S. Haroche. Trapping and coherent manipulation of a rydberg atom on a microfabricated device: a proposal. *Eur. Phys. J. D*, 35:43–57, 2005.

[21] D. Cano, B. Kasch, H. Hattermann, R. Kleiner, C. Zimmermann, D. Kölle, and J. Fortagh. Meissner eect in superconducting microtraps. *Phys. Rev. Lett*, 101:1830, 2008.

[22] D. Cano, B. Kasch, H. Hattermann, D. Kölle, R. Kleiner, C. Zimmermann, and J. Fortagh. Impact of the meissner eect on magnetic microtraps for neutral atoms near superconducting thin films. *Phys. Rev. A*, 77:063408, 2008.

[23] Ch. Hufnagel, T. Mukai, and F. Shimizu. Stability of a superconductive atom chip with persistent current. *Phys. Rev. A*, 79:05364, 2009.

[24] A. Emmert, A. Lupascu, G. Nogues, M. Brune, J.-M. Raimond, and S. Haroche. Measurement of the trapping lifetime close to a cold metallic surface on a cryogenic atomchip. *Eur. Phys. J. D*, 51:173–177, 2009.

[25] B. Kasch, H. Hattermann, D. Cano, T. E. Judd, S. Scheel, C. Zimmermann, R. Kleiner, D. Kölle, and J. Fortagh. Cold atoms near superconductors: Atomic spin coherence beyond the johnson noise limit. *New Jour. of Phys.*, 12:065024, 2009. arXiv:0906.1369v2 [cond-mat.supr-con] 4 Dec 2009.

Bibliography

[26] C. Roux, A. Emmert, A. Lupascu, T. Nirrengarten, G. Nogues, M. Brune, J.-M. Raimond, and S. Haroche. Bose-einstein condensation on a superconducting atom chip. *Euro. Phys. Lett*, 81:56004, 2008.

[27] T. Müller, X. Wu, A. Mohan, A. Eyvazov, Y. Wu, and R. Dumke. Towards a guided atom interferometer based on a superconducting atom chip. *New Jour. of Phys.*, 10:073006, 2008.

[28] T. Mueller, B. Zhang, R. Fermani, K. S. Chan, Z. W. Wang, C. B. Zhang, M. J. Lim, and R. Dumke. Trapping of ultra-cold atoms with the magnetic field of vortices in a thin-film superconducting micro-structure. *New Jour. of Phys.*, 12:043016, 2010.

[29] R Fermani, T. Mueller, B. Zhang, M. J. Lim, and R. Dumke. Heating rate and spin flip lifetime due to near-field noise in layered superconducting atom chips. *J. Phys. B: At. Mol. Opt. Phys.*, 43:095002, 2010.

[30] L. Tian, P. Rabl, R. Blatt, and P. Zoller. Interfacing quantum-optical and solid-state qubits. *Phys. Rev. Lett.*, 92(24):247902–1, 2004.

[31] J. Verdu, H. Zoubi, Ch. Koller, J. Majer, H. Ritsch, and J. Schmiedmayer. Strong magnetic coupling of an ultracold gas to a superconducting waveguide cavity. *Phys. Rev. Lett.*, 103:436, 2009.

[32] A. S. Soerensen, C. H. van der Wal, L. I. Childress, and M. D. Lukin. Capacitive coupling of atomic systems to mesoscopic conductors. *Phys. Rev. Lett.*, 92:636, 2004.

[33] D. Petrosyan, G. Bensky, G. Kurizki, I. Mazets, J. Majer, and J. Schmiedmayer. Reversible state transfer between superconducting qubits and atomic ensembles. *Phys. Rev. A*, 79:040304(R), 2009.

[34] D. Petrosyan and M. Fleischhauer. Quantum information processing with single photons and atomic ensembles in microwave coplanar waveguide resonators. *Phys. Rev. Lett.*, 100:1705, 2008.

[35] L. M. Duan, J. I. Cirac M. D. Lukin, and P. Zoller. Long-distance quantum communication with atomic ensembles and linear optics. *Nature*, 414:413, 2001.

[36] P. Bushev, A. K. Feofanov, H. Rotzinger, I. Protopopov, J. H. Cole, G. Fischer, A. Lukashenko, and A. V. Ustinov. Rare earth spin ensemble magnetically coupled to a superconducting resonator. *arXiv:1102.3841v1 [cond-mat.mes-hall]*, v1:6, 2011.

[37] D. I. Schuster, A. P. Sears, E. Ginossar, L. DiCarlo, L. Frunzio, J. J. L. Morton, H. Wu, G. A. D. Briggs, B. B. Buckley, D. D. Awschalom, and R. J. Schoelkopf. High-cooperativity coupling of electron-spin ensembles to superconducting cavities. *Phys. Rev. Lett.*, 105:140501, 2010.

[38] Y. Kubo, F. R. Ong, P. Bertet, D. Vion, V. Jacques, D. Zheng, A. Dréau, J.-F. Roch, A. Aueves, F. Jelezko, J. Wrachtrup, M. F. Barthe, P. Bergonzo, and D. Esteve. Strong coupling of a spin ensemble to a superconducting resonator. *Phys. Rev. Lett.*, 105:140502, 2010.

Bibliography

[39] P. Treutlein, P. Hommelho, T. Steinmetz, T. W. Hänsch, and J. Reichel. Coherence in microchip traps. *Phys. Rev. Lett.*, 92:2030, 2004.

[40] P. A. Willems and K. G. Libbrecht. Creating long-lived neutral-atom traps in a cryogenic environment. *Phys. Rev. A*, 51:1403, 1995.

[41] M. H. Anderson, J. R. Ensher, M. R. Matthews, C. E. Wieman, and E. A. Cornell. Observation of bose-einstein condensation in a dilute atomic vapor. *Science*, 269:198, 1995.

[42] K. B. Davis, M. O. Mewes, M. R. Andrews, N. J. van Druten, D. S. Durfee, D. M. Kurn, and W. Ketterle. Bose-einstein condensation in a gas of sodium atoms. *Phys. Rev. Lett.*, 75:3969, 1995.

[43] T. Weber, J. Herbig, M. Mark, H.C. Nägerl, and R. Grimm. Bose-einstein condensation of cesium. *Science*, 299:232–235, 2003.

[44] G. Modugno, G. Ferrari, G. Roati, R. J. Brecha, A. Simoni, and M. Inguscio. Bose-einstein condensation of potassium atoms by sympathetic cooling. *Science*, 294:1320–1322, 2001.

[45] C. C. Bradley, C. A. Sackett, and R. G. Hulet. Bose-einstein condensation of lithium: Observation of limited condensate number. *Phys. Rev. Lett.*, 78:985, 1997.

[46] Z. Hadzibabic, C. A. Stan, K. Dieckmann, S. Gupta, M. W. Zwierlein, A. Goerlitz, and W. Ketterle. Two-species mixture of quantum degenerate bose and fermi gases. *Phys. Rev. Lett.*, 88:160401, 2002.

[47] M. D. Ray T. E. Barrett, S. W. Dapore-Schwartz and G. P. Lafyatis. Slowing atoms with sigma- polarized light. *Phys. Rev. Lett.*, 67:3483, 1991.

[48] W. F. van Dorp and C. W. Hagen. A critical literature review of focused electron beam induced deposition. *Journal of Applied Physics*, 104:081301, 2008.

[49] C. Boothroyd. Microanalysis in the electron microscope. Technical report, Center for electron nanoscopy, DTU Denmark, 2009.

[50] B. Ziaja, R. A. London, and J. Hajduc. Ionization by impact electrons in solids: Electron mean free path fitted over a wide energy range. *Journal of Applied Physics*, 99:033514, 2006.

[51] K. Kanaya and S. Okayama. Penetration and energy-loss theory of electrons in solid targets. *J. Phys. D: Appl. Phys.*, 5:43–58, 1972.

[52] J. C. Ashley and V. E. Anderson. Interaction of low energy electrons with silicon dioxide. *J. Elect. Spectrosc.*, 24:127, 1981.

[53] T. E. Everhart and P. H. Ho. Determination of kilovolt electron energy dissipation vs penetration distance in solid materials. *J. Appl. Phys.*, 42:5837, 1971.

[54] H. Niedrig. Electron backscattering from thin films. *Journal of Applied Physics*, 53:R15, 1982.

[55] G. K. Wertheim, D. M. Rie, N. V. Smith, and P. H. Citrin. Electron mean free path in the alkali metals. *Phys. Rev. B*, 46:1955, 1992.

[56] H. Bethe. Leipzig. *Ann. Physik.*, 5:325, 1930.

[57] Y.-K. Kim, J. Migdalek, W. Siegel, and J. Bieron. Electron-impact ionization cross section of rubidium. *Phys. Rev. A*, 57:246, 1998.

[58] Y. Sakai, A. Haga, S. Sugita, S. Kita, S.-I. Tanaka, F. Okuyama, and N. Kobayashi. Electron gun using carbon-nanofiber field emitter. *Rev. Sci. Instr.*, 78:013305, 2007.

[59] J. Liu, Z. Zhang, Y. Zhao, X. Su, S. Liu, and E. Wang. Tuning the field-emission properties of tungsten oxide nanorods. *small*, 1:310, 2005.

[60] S. Yamamoto. Fundamental physics of vacuum electron sources. *Rep. Prog. Phys.*, 69:181, 2006.

[61] R. H. Fowler and L. Nordheim. Electron emission in intense electric fields. *Proc. R. Soc. Lond. A*, 119:173–181, 1928.

[62] J. C. Wiesner and T. E. Everhart. Point-cathode electron sources - electron optics of the initial diode region. *Journal of Applied Physics*, 44:2140, 1973.

[63] T. Wang, C. Reece, and R. Sundelin. Field emission studies from nb surfaces relevant to srf vavities. In *IEEE Proceedings of the 2003 Particle Accelerator Conference*, 2003.

[64] A.V. Crewe, D. N. Eggenberger, J. Wall, and L.M. Welter. Electron gun using a field emission source. *Rev. Sci. Instr.*, 39:546, 1968.

[65] K. S. Yeong and J. T. L. Thong. Life cycle of a tungsten cold field emitter. *Journal of Applied Physics*, 99:104903, 2006.

[66] Jackson. *Jackson - Klassische Elektrodynamik*. 3. überarb. Aufl.,Berlin ISBN 3-11-016502-3, 2002.

[67] T. Hänsch and A. Schawlow. Cooling of gases by laser radiation. *Optical Communications*, 13:68, 1975.

[68] S. Chu. Nobel lecture: The manipulation of neutral particles. *Rev. Mod. Phys.*, 70:685, 1998.

[69] W. D. Phillips. Nobel lecture: Laser cooling and trapping of neutral atoms. *Rev. Mod. Phys.*, 70:721, 1998.

[70] C. Cohen-Tannoudji. Manipulating atoms with photons. *Rev. Mod. Phys*, 70:707, 1998.

[71] C. V. Heer. Feasibility of containment of quantum magnetic dipoles. *Rev. Sci. Instr.*, 34:532, 1963.

[72] A. L. Migdall, J. V. Prodan, T. H. Bergeman, H. J. Metcalf, and W. D. Phillips. First observation of magnetically trapped neutral atoms. *Phys. Rev. Lett.*, 54:2596–2599, 1985.

Bibliography

[73] D. E. Pritchard. Cooling neutral atoms in a magnetic trap for precision spectroscopy. *Phys. Rev. Lett.*, 51:1336, 1983.

[74] N. Masuhara, J. M. Doyle, J. C. Sandburg, D. Kleppner, T. J. Greytak, H. F. Hess, and G. P. Kochansky. Evaporative cooling of spin-polarized atomic hydrogen. *Phys. Rev. Lett*, 61:935, 1988.

[75] C. J. Pethick and H. Smith. *Bose-Einstein Condensation in Dilute Gases*. Cambridge University Press, 2002.

[76] E. A. Hinds and I. G. Hughes. Magnetic atom optics: mirrors, guides, traps, and chips for atoms. *Journal of Physics D*, 32:119, 1999.

[77] E. U.; G. H. Shortley Condon. *The Theory of Atomic Spectra*. Cambridge University Press. ISBN 0-521-09209-4., 1935.

[78] W. Gerlach and O. Stern. Der experimentelle nachweis der richtungsquantelung im magnetfeld. *Zeitschrift für Physik*, 9:349, 1922.

[79] P. O. Schmidt, S. Hensler, J. Werner, A. Görlitz T. Binhammer, and T. Pfau. Continuous loading of cold atoms into a ioe-pritchard magnetic trap. *Journal of Optics B: Quantum and Semiclassical Optics*, 5:170, 2003.

[80] S. Earnshaw. On the nature of the molecular forces which regulate the constitution of the luminiferous ether. *Trans. Camb. Phil. Soc.*, 7:97, 1842.

[81] E. Raab, M. Prentiss, A. Cable, S. Chu, and D. Pritchard. Trapping of neutral-sodium atoms with radiation pressure. *Phys. Rev. Lett.*, 59:2631, 1987.

[82] Christopher J. Foot. *Atomic Physics*. Oxford University Press, 2005.

[83] H. Metcalf. Magneto-optical trapping and its application to helium metastables. *J. Opt. Soc. Am. B*, 6:2206, 1989.

[84] H. J. Metcalf. *Laser Cooling and Trapping*. Springer Verlag, Heidelberg Berlin New York, 1999.

[85] S. Chu and C. Wieman. Laser cooling and trapping of atoms. *J. Opt. Soc. Am. B*, 6:1961, 1989.

[86] P. Lett, R. Watts, C. Westbrook, W. Phillips, P. Gould, and H. Metcalf. Observation of atoms laser cooled below the doppler limit. *Phys. Rev. Lett.*, 61:169, 1988.

[87] J. Dalibard and C. Cohen-Tannoudji. Dressed atom approach to atomic motion in laser light: the dipole force revisited. *J. Opt. Soc. Am. B*, 2:1707, 1985.

[88] J. Dalibard and C. Cohen-Tannoudji. Laser cooling below the doppler limit by polarization gradients, simple theoretical models. *J. Opt. Soc. Am. B*, 6:2023, 1989.

[89] M. Weidemüller and C. Zimmermann. *Interactions in Ultracold Gases*. Wiley-VCH, Berlin, 2003.

[90] T. Walker, D. Sesko, and C. Wieman. Collective behavior of optically trapped neutral atoms. *Phys. Rev. Lett.*, 64:408–411, 1990.

[91] M. Prentiss, A. Cable, J. E. Bjorkholm, S. Chu, E. L. Raab, and D. E. Pritchard. Atomic-density-dependent losses in an optical trap. *Optics Letters*, 13(6):452–454, 1988.

[92] J. Schmiedmayer. Guiding and trapping a neutral atom on a wire. *Phys. Rev. A*, 52:52, 1995.

[93] W. Petrich, M. H. Anderson, J. R. Ensher, and E. A. Cornell. Stable, tightly confining magnetic trap for evaporative cooling of neutral atoms. *Phys. Rev. Lett.*, 74:3352, 1995.

[94] M. S. Ioe and R. I. Sobolev. Confinement of a plasma in a trap formed by a combined magnetic field. *Plasma Physics (Journal of Nuclear Energy Part C)*, 7:501, 1965.

[95] T. Bergeman, G. Erez, and H. J. Metcalf. Magnetostatic trapping fields for neutral atoms. *Phys. Rev. A*, 1987:1535, 35.

[96] V. S. Bagnato, G. P. Lafyatis, A. G. Martin, E. L. Raab, R. N. Ahmad-Bitar, and D. E. Pritchard. Continuous stopping and trapping of neutral atoms. *Phys. Rev. Lett.*, 58:2194, 1987.

[97] C. C. Bradley. Evidence of bose-einstein condensation in an atomic gas with attractive interactions. *Phys. Rev. Lett.*, 75:1687, 1995.

[98] M.-O. Mewes, M. R. Andrews, N. J. van Druten, D. M. Kurn, D. S. Durfee, and W. Ketterle. Bose-einstein condensation in a tightly confining dc magnetic trap. *Phys. Rev. Lett.*, 77:416–419, 1996.

[99] Hau L., in *Photonic, Electronic, and Atomic Collisions, Proceedings of the XX IC-PEAC, Vienna, Austria, 1997*, edited by F. Aumayr and H. P. Winter (World Scientific), 1998.

[100] C. J. Myatt, E. A. Burt, R. W. Ghrist, E. A. Cornell, and C. E. Wieman. Production of two overlapping bose-einstein condensates by sympathetic cooling. *Phys. Rev. Lett*, 78:586, 1997.

[101] R.G. Dall and A.G. Truscott. Bose̊einstein condensation of metastable helium in a bi-planar quadrupole ioe configuration trap. *Optics Communications*, 270:255–261, 2007.

[102] CHEN Shuai, ZHOU Xiao-Ji, YANG Fan, XIA Lin, WANG Yi-Qiu, and CHEN Xu-Zong. Optimization of the loading process of the quic magnetic trap for the experiment of bose̊einstein condensation. *Chin. Phys. Lett*, 21(11):2227, 2004.

[103] D. Cano, B. Kasch, H. Hattermann, D. Kölle, R. Kleiner, C. Zimmermann, and J. Fortagh. Impact of the meissner eect on magnetic microtraps for neutral atoms near superconducting thin films. *Phys. Rev. A*, 77:063408, 2008.

[104] V. Sokolovsky, L. Prigozhin, and V. Dikovsky. Meissner transport current in flat films of arbitrary shape and a magnetic trap for cold atoms. *Supercond. Sci. Technol.*, 23:065003, 2010.

[105] V. Dikovsky, V. Sokolovsky, B. Zhang, C. Henkel, and R. Folman. Superconducting atom chips: advantages and challenges. *Eur. Phys. J. D*, 51:247–259, 2009.

[106] S. Schneider. *Bose-Einstein Kondensation in einer magnetischen Z-Falle*. PhD thesis, Ruprecht-Karls-Universität, Heidelberg, 2003.

[107] D. Cassettari A. Haase, and, B. Hessmo, and J.Schmiedmayer. Trapping neutral atoms with a wire. *Phys. Rev. A*, 64:043405, 2001.

[108] E. Majorana. Atomi orientati in campo magnetico variabile. *Il Nuovo Cimento*, 9(2):43–50, 1924-1942.

[109] D. M Brink and C. V Sukumar. Majorana spin-flip transitions in a magnetic trap. *Phys. Rev. A*, 74:035401, 2006.

[110] S. Gov, S. Shtrikman, and H. Thomas. Magnetic trapping of neutral particles: Classical and quantum-mechanical study of a ioe-pritchard type trap. *Journal of Applied Physics*, 87:3989, 2000.

[111] S. Bali, K.M OŠHara, M.E. Gehm, S.R. Granade, and J.E. Thomas. Quantumdiractive background gas collisions in atom-trap heating and loss. *Phys. Rev. A*, 60:R29, 1999.

[112] H.M.J.M. Boesten, A.J. Moerdijk, and B.J. Verhaar. Dipolar decay in two recent bose-einstein experiments. *Phys. Rev. A*, 54:R29, 1996.

[113] D.W. Snoke and J.P. Wolfe. Population dynamics of a bose gas near saturation. *Phys. Rev. B*, 39:4030, 1989.

[114] D. W. Snoke, W. W. Ruhle, Y. G. Lu, and E. Bauer. Evolution of a nonthermal electron energy distribution in gaas. *Phys. Rev. B*, 45:10979, 1992.

[115] D.J. Heinzen. *Ultra cold atomic interactions, Proceedings of the International School of Physics: Enrico Fermi*. IOS Press, 1999.

[116] T. A. Savard, K. M. OŠHara, and J. E. Thomas. Laser-noise-induced heating in far-o resonance optical traps. *Phys. Rev. A*, 56:1095–1098, 1997.

[117] W. Ketterle. *Making, Probing and Understanding Bose-Einstein Condensation, Proceedings of the International School of Physics: Enrico Fermi*. IOS Press, 1999.

[118] C. Henkel, P. Krüger, R. Folman, and J. Schmiedmayer. Fundamental limits for coherent manipulation on atom chips. *Appl. Phys. B*, 76:173, 2003.

[119] M.E. Gehm, K.M OŠHara, T.A Savard, and J.E. Thomas. Dynamics of noise-induced heating in atom traps. *Phys. Rev. A*, 58:3914, 1998.

[120] A. Burchianti, A. Bogi, C. Marinelli, E. Mariotti, and L. Moi. Light-induced atomic desorption and related phenomena. *Phys. Scripta.*, T135:014012, 2009.

[121] C. Marinelli, A. Burchianti, A. Bogi, F. Della Valle, G. Bevilacqua, E. Mariotti, S. Veronesi, and L. Moi. Desorption of rb and cs from pdms induced by non resonant light scattering liad. *Eur. Phys. J. D*, 37:319, 2006.

[122] B. P. Anderson and M. A. Kasevich. Loading a vapor-cell magneto-optic trap using light-induced atom desorption. *Phys. Rev. A*, 63:023404, 2001.

[123] SAES Getters. Alkali metal dispenser datasheed. Technical report, S.p.A 20151 Milano, Italy, 2003.

[124] C. Zimmermann J. Fortagh, A. Grossmann and T. W. Hänsch. Fast loading of a magneto-optical trap from a pulsed thermal source. *Journal of Applied Physics*, 84:6499, 1998.

[125] T. Hong, A. V. Gorshkov, D. Patterson, A. S. Zibrov, J. M. Doyle, M D. Lukin, and Mara G. Prentiss. Realization of coherent optically dense media via bu er-gas cooling. *Phys. Rev. A*, 79:013806, 2009.

[126] P. F. Gri n, K. J. Weatherill, and C. S. Adams. Fast switching of alkali atom dispensers using laser-induced heating. *Rev. Sci. Instr.*, 76:093102, 2005.

[127] S. E. Maxwell, N. Brahms, R. deCarvalho, D. R. Glenn, J. S. Helton, S.V. Nguyen, D. Patterson, J. Petricka, D. DeMille, , and J. M. Doyle. High-flux beam source for cold, slow atoms or molecules. *Phys. Rev. Lett.*, 95:173201, 2005.

[128] V.N. Ageev, Yu.A. Kuznetsov, and N.D. Potekhina. Electron-stimulated desorption of alkali metal and barium atoms from an oxidized tungsten surface. *Surface Science 367 (1996)*, 376:113–127, 1996.

[129] V. N. Ageev, Yu.A. Kuznetsov, and T.E. Madey. Electron-stimulated desorption of potassium and cesium atoms from oxidized molybdenum surfaces. *Surface Science*, 390:146–151, 1997.

[130] V. N. Ageev, Yu. A. Kuznetsov, and T.E. Madey. Electron-stimulated desorption of lithium atoms from oxygen-covered molybdenum surfaces. *Surface Science*, 451:153–159, 2000.

[131] V. N. Ageev and Yu. A. Kuznetsov. Electron-stimulated desorption of sodium atoms from an oxidized molybdenum surface. *Phys. Rev. B*, 58:2248, 1998.

[132] P. A. Redhead. The first 50 years of electron stimulated desorption (1918-1968). *Vacuum*, 4(6):585–596, 1997.

[133] V. N. Ageev. Desorption induced by electronic transitions. *Prog. Surf. Sci.*, 47:55, 1994.

[134] T. E. Madey. History of desorption induced by electronic transitions. *Surface Science*, 299-300:824–836, 1994.

[135] P. J. Feibelman and M. L. Knotek. Reinterpretation of electron-stimulated desorption data from chemisorption systems. *Phys. Rev. B*, 18:6531–6540, 1978.

[136] V.N. Ageev, Y. A. Kuznetsova, B.V. Yakshinskii, and T.E. Madey. Electron stimulated desorption of alkali metal ions and atoms: Local surface field relaxation. *Nuclear Instruments and Methods in Physics Research B*, 101:69–72, 1995.

[137] M. L. Knotek, V. O Jones, and V. Rehn. Photon-stimulated desorption of ions. *Phys. Rev. Lett.*, 43(4):300–303, 1979.

Bibliography

[138] P. R. Antoniewicz. Model for electron- and photon-stimulated desorption. *Phys. Rev. B*, 21(9):3811–3815, 1980.

[139] B. V. Yakshinskiy and T. E. Madey. Photon-stimulated desorption as a substantial source of sodium in the lunar atmosphere. *Nature*, 400:642–644, 1999.

[140] D. A. Steck. Rubidium 87 d line data. Technical report, Oregon Center for Optics and Department of Physics, University of Oregon, 2008.

[141] D. A. Steck. Cesium d line data. Technical report, Theoretical Division (T-8), MS B285 Los Alamos National Laboratory, Los Alamos, NM 87545, 2003.

[142] D. A. Steck. Sodium d line data. Technical report, Theoretical Division (T-8), MS B285 Los Alamos National Laboratory Los Alamos, NM 87545, 2003.

[143] F. Geiger, C. A. Busse, and R. I. Loehrke. The vapor pressure of indium, silver, gallium, copper, tin, and gold between 0.1 and 3.0 bar. *International Journal of Thermophysics*, 8(4):425–436, 1987.

[144] W. T. Hicks. Evaluation of vapor-pressure data for mercury, lithium, sodium, and potassium. *The Journal of Chemical Physics*, 38(8):1873, 1963.

[145] T. P. Lin. Estimation of temperature rise in electron beam heating of thin films. *IBM Journal of Research and Development*, 11:527–536, 1967.

[146] I. Langmuir. The vapour pressure of metallic tungsten. *Phys. Rev.*, 11:329, 1913.

[147] D. Drouin, A. R. Couture, D. Joly, X. Tastet, V. Aimez, and R. Gauvin. Casino v2.42 - a fast and easy-to-use modeling tool for scanning electron microscopy and microanalysis users. *Scanning*, 29:92–101, 2007.

[148] J. W. Motz, H. Olsen, and H. W. Koch. Electron scattering without atomic or nulcear excitation. *Rev. Mod. Phys*, 1964:881–928, 36.

[149] R. Browning, T. Z. Li, B. Chui, J. Ye, R. F. W. Pease, Z. Czyzewski, and D. C. Joy. Low-energy electron/atom elastic scattering cross sections from 0.1Ű30 kev. *Scanning*, 17(4):250–253, 1995.

[150] O. S. Heavens. Evaporation of metals by electron bombardment. *J. Sci. Instr.*, 36:95–95, 1959.

[151] A. Powell, P. Minson, G. Trapaga, and U. Pal. Mathematical modeling of vapor-plume focusing in electron-beam evaporation. *Metallurgical and Materials Transactions A*, 32A:1, 2001.

[152] M. Kndusen. Das cosinusgesetz in der kinetischen gastheorie. *Ann. Physik.*, 48:471, 1930.

[153] E. B. Graper. Distribution and apparent source geometry of electron-beam-heated evaporation. *J. Vac. Sci. and Techn.*, 10:100, 1973.

[154] R. Amsüss. Development of a source of ultracold atoms for cryogenic environments. Master's thesis, Technische Universität Wien, Atominstitut, 2008.

[155] B. L. Rogers, J. G. Shapter, W. M. Skinner, and K. Gascoigne. A method for production of cheap, reliable ptÜir tips. *Rev. Sci. Instr.*, 71:1702, 2000.

[156] T. Suzuki K. Kuroda. High current e ciency accelerating lens system of field emission scanning electron microscope. *Journal of Applied Physics*, 46:454, 76.

[157] D.W. Tuggle and L.W. Swanson. Emission characteristics of the zro / w thermal field electron source. *J. Vac. Sci. Technol. B*, 3:220, 1985.

[158] K. Kuroda, H. Ebisui, and T. Suzuki. Three-anode acceleration lens system for the field emission scanning electron microscope. *Journal of Applied Physics*, 45:2336, 1974.

[159] K. Kuroda and T. Suzuki. Analysis of accelerating lens system for field emission sem. *Journal of Applied Physics*, 45:1436, 1974.

[160] T. Gericke, P. Würtz, D. Reitz, T. Langen, and H. Ott. High-resolution scanning electron microscopy of an ultracold quantum gas. *Nature. Phys.*, 4:949, 2008.

[161] P. Würtz, T. Gericke, A. Vogler, and H. Ott. Ultracold atoms as a target: absolute scattering cross-section measurements. *New Journal of Physics*, 12:065033, 2010.

[162] R. S. Schappe, P. Feng, L. W. Anderson, C. C. Lin, and T. Walker. Electron collision cross-sections measured with the use of a magneto-optical trap. *Euro Phys. Lett.*, 29(6):439–44, 1995.

[163] R. S. Schappe, T. Walker, L. W. Anderson, and Chun C. Lin. Absolute electron-impact ionization cross section measurements using a magneto-optical trap. *Phys. Rev. Lett*, 76:4328, 1996.

[164] M. Lukomski, J. A. MacAskill, D. P. Seccombe, C. McGrath, S. Sutton, J. Teeuwen, W. Kedzierski, T. J. Reddish, J. W. McConkey, and W A van Wijngaarden5. New measurements of absolute total cross sections for electron impact on caesium using a magneto-optical trap. *J. Phys. B: At. Mol. Opt. Phys.*, 38:3535–3545, 2005.

[165] M. L. Keeler, L.W. Anderson, and Chun C. Lin. Electron-impact ionization cross section measurements out of the excited state of rubidium. *Phys. Rev. Lett.*, 85:3353, 2000.

[166] R. H. McFarland. Gryzinski electron-impact ionization cross-section computations for the alkali metals. *Phys. Rev.*, 139:A40, 1965.

[167] E. M. Mueller and T. T. Tsong. *Field ion microscopy*. American Elsevier Publishing Company, New York, 1969.

[168] T. Hirayama, A. Hayama, T. Adachi, I. Arakawa, and M. Sakurai. Desorption of excimers from the surface of solid ne by low-energy electron or photon impact. *Phys. Rev. B*, 63:075407, 2001.

[169] S. Casalbuoni, A. Grau, M. Hagelstein, R. Rossmanith, F. Zimmermann, B. Kostka, E. Mashkina, E. Steens, A. Bernhard, D. Wollmann, and T. Baumbach. Beam heat load and pressure rise in a cold vacuum chamber. *Phys. Rev. Special Topics - Accelerators and Beams*, 10:093202, 2007.

Bibliography

[170] H. Tratnik. *Electron Stimulated Desorption of Condensed Gases on Cryogenic Surfaces*. PhD thesis, Technischen Universität Wien, Fakultät für Technische Naturwissenschaften und Informatik, 2005.

[171] V.N. Ageev, Y. A. Kuznetsova, and T.E. Madey. Temperature dependences in electron-stimulated desorption of neutral europium. *Journal of Electron Spectroscopy and Related Phenomena*, 128:223–229, 2003.

[172] J. Gomez-Gonia and A. G. Mathewson. Temperature dependence of the electron induced gas desorption yields on stainless steel, copper, and aluminum. *J. Vac. Sci. Technol.*, 15(6):3093–3103, 1997.

[173] V. N. Ageev and Yu. A. Kuznetsov. The yield of cesium atoms in electron-stimulated desorption from germanium-covered tungsten. *Technical Physics Letters*, 31(3):249–251, 2005.

[174] T. L. Gustavson, A. P. Chikkatur, A. E. Leanhardt, A. Görlitz, S. Gupta, D. E. Pritchard, and W. Ketterle. Transport of bose-einstein condensates with optical tweezers. *Phys. Rev. Lett.*, 88:020401–1, 2002.

[175] M. Greiner, I. Bloch, T. W. Hansch, and T. Esslinger. Magnetic transport of trapped cold atoms over a large distance. *Phys. Rev. A*, 63:031401–1, 2001.

[176] H.J. Lewandowski, D.M. Harber, D.L. Whitaker, and E.A. Cornell. Simplified system for creating a bose-einstein condensate. *J. Low Temp. Phys.*, 132:309, 2003.

[177] K. Nakagawa, Y. Suzuki, and J. B. Kim M. Horikoshi. Simple and e cient magnetic transport of cold atoms using moving coils for the production of boseŨeinstein condensation. *Appl. Phys. B*, 81:791–794, 2005.

[178] N. Lippok. A magnetic transport for ultracold atoms. Master's thesis, Ruperto-Carola University of Heidelberg, 2008.

[179] M. Wilzbach. *Single atom detection on an atom chip with integrated optics*. PhD thesis, Ruperto-Carola University of Heidelberg, 2007.

[180] U. Schünemann, H. Engler, R. Grimm, M. Weidemüller, and M. Zielonkowski. Simple scheme for tunable frequency oset locking of two lasers. *Rev. Sci. Instr.*, 70:242–243, 1999.

[181] S. Kraft, A. Deninger, C. Trück, J. Fortagh, F. Lison, and C. Zimmermann. Rubidium spectroscopy at 778-780 nm with a distributed feedback laser diode. *Laser Phys. Let.*, 2:71–76, 2004.

[182] Jack W. Ekin. *Experimental techniques for low-temperature measurements. Cryostat design, material properties and superconducting critical current testing*. Oxford University Press, 2006.

[183] G. Ventura and L. Risegari. *The Art of Cryogenics, Low-Temperature Experimental Techniques*. Elsevier Ltd, 2008.

[184] Frank Pobell. *Matter and Methods at Low Temperatures*. Springer Verlag, Heidelberg Berlin, 2007.

[185] K. Uhlig and W. Hehn. He3/he4 dilution refrigerator precooled by giord-mcmahon refrigerator. *Cryogenics*, 37:279, 1997.

[186] K. Uhlig. He3/he4 dilution refrigerator precooled by giord-mcmahon cooler ii. measurements of the vibrational heat leak. *Cryogenics*, 42:569, 2002.

[187] Y. Ikushima, R. Li, T. Tomaru, N. Sato, T. Suzuki, T. Haruyama, T. Shintomi, and A. Yamamoto. Ultra-low-vibration pulse-tube cryocooler system - cooling capacity and vibration. *Cryogenics*, 48:406–412, 2008.

[188] W. Ketterle, D.S. Durfee, and D.M. Stamper-Kurn. Making, probing and understanding bose-einstein condensates. *arXiv:cond-mat/9904034v2*, 2:87, 1999.

[189] T. Esslinger, I. Bloch, and T. W. Hänsch. Bose-einstein condensation in a quadrupole-ioe-configuration trap. *Phys. Rev. A*, 58:R2664–R2667, 1998.

[190] W. Meissner and R. Ochsenfeld. Ein neuer ekt bei eintritt der supraleitfähigkeit. *Naturwissenschaften*, 21:787–788, 1933.

[191] J. G. Hust. Thermal anchoring of wires in cryogenic apparatus. *Rev. Sci. Instr.*, 41:622, 1970.

[192] T. P. Meyrath. Electromagnet design basics for cold atom experiments. Technical report, Atom Optics Laboratory, Center for Nonlinear Dynamics, University of Texas at Austin, 2003 (2004).

[193] Y. Miura. An optical window with thermal contraction free seal for low temperature use. *Cryogenics*, 22:374, 1982.

[194] D. Celik and S.W. Van Sciver. Large size optical windows for superfluid helium applications. *Cryogenics*, 42:547–549, 2002.

[195] P. Gebhardt, R. Helbig, and K. Huemmer. An optical window for metal cryostats. *Cryogenics*, 17:735, 1976.

[196] G. B. Gorlin. The vacuum sealing of optical windows for operation at low temperature. *Cryogenics*, 14:407, 1974.

[197] E. Beckman and R. Rass. Window seals for metal cryostats. *Cryogenics*, 11:147, 1971.

[198] P. Loose, U. Waas, and M. Wohlecke. Improved method for sealing windows in metal cryostats. *Cryogenics*, 14:470, 1974.

[199] L. F. Mollenauer, C. D. Grandt, W. B. Grant, and H. Panepucci. Strain-free, fused silica optical windows for a metal dewar. *Rev. Sci. Instr.*, 39:1683290, 1958.

[200] J. Fesmire, S. Augustynowicz, and C. Darve. Performance characterization of perforated multilayer insulation blankets. Technical report, NASA Kennedy Space Center, FL 32899, USA, 2002.

Bibliography

[201] B. Zhang, R. Fermani, T. Müller, M. J. Lim, and R. Dumke. Design of magnetic traps for neutral atoms with vortices in type-ii superconducting micro-structures. *arXiv:1004.0064v1*, 1:11, 2010. arXiv:1004.0064v1 [physics.atom-ph] 1 Apr 2010.

[202] K. Davis, M.-O. Mewes, M. Joe, M. Andrews, and W. Ketterle. Evaporative cooling of sodium atoms. *Phys. Rev. Lett.*, 74:5202, 1995.

[203] H. Yamada, N. Harada, Kanayama, Nakagawa, H. Yamasaki, and T. Hamajima. Enhancement of transport critical current density of epitaxial nb film by lithography. *Physica C*, 433:65–69, 2005.

[204] N. Debernardi, M. P. Reijnders, W. J. Engelen, T. T. J. Clevis, P. H. A. Mutsaers, O. J. Luiten, and E. J. D. Vredenbregt. Measurement of the temperature of an ultracold ion source using time-dependent electric fields. *arXiv:1011.2369v4 [physics.atom-ph]*, 4:3, 2011.

[205] G. Nogues, C. Roux, T. Nirrengarten, A. Lupascu, A. Emmert, M. Brune, J.-M. Raimond, S. Haroche, B. Placais, and J.-J. Greet. Eect of vortices on the spin-flip lifetime of atoms in superconducting atom-chips. *Euro. Phys. Lett*, 87:13002, 2009.

[206] K. Henschel, J. Majer, J. Schmiedmayer, and Helmut Ritsch. Cavity qed with an ultracold ensemble on a chip: Prospects for strong magnetic coupling at finite temperatures. *Phys. Rev. A*, 82:033810, 2010.

[207] P. Rabl, D. DeMille, J. M. Doyle, M. D. Lukin, R. J. Schoelkopf, and P. Zoller. Hybrid quantum processors: Molecular ensembles as quantum memory for solid state circuits. *Phys. Rev. Lett.*, 97:033003, 2006.

[208] A. Imamoglu. Cavity qed based on collective magnetic dipole coupling: Spin ensembles as hybrid two-level systems. *Phys. Rev. Lett.*, 102:083602, 2009.

[209] E. T. Jaynes and F. W. Cummings. Comparison of quantum and semiclassical radiation theories with application to the beam maser. *Proceedings of the IEEE*, 51(1):89–109, 1963.

[210] R. J. Schoelkopf and S. M. Girvin. Wiring up quantum systems. *Nature*, 451:664–669, 2008.

[211] S. Haroche and J. M. Raimond. *Exploring the Quantum: Atoms, Cavities, and Photons*. Oxford University Press, 2006.

[212] M. Devoret, S. Girvin, and R. Schoelkopf. Circuit-qed: How strong can the coupling between a josephson junction atom and a transmission line resonator be? *Ann. Phys.*, 16:767Ű779, 2007.

[213] M. G. Raizen, R. J. Thompson, R. J. Brecha, H. J. Kimble, and H. J. Carmichael. Normal-mode splitting and linewidth averaging for two-state atoms in an optical cavity. *Phys. Rev. Lett.*, 63:240–243, 1989.

[214] J. M. Fink, R. Bianchetti, M. Baur, M. Göppl, L. Steen, S. Filipp, P. J. Leek, A. Blais, and A. Wallra. Dressed collective qubit states and the tavis-cummings model in circuit qed. *Phys. Rev. Lett.*, 103:083601, 2009.

[215] Y. Kaluzny, P. Goy, M. Gross, J. M. Raimond, and S. Haroche. Observation of self-induced rabi oscillations in two-level atoms excited inside a resonant cavity: The ringing regime of superradiance. *Phys. Rev. Lett.*, 51:1175–1178, 1983.

[216] M. Tavis and F. W. Cummings. Exact solution for an n-molecule-radiation-field hamiltonian. *Phys. Rev.*, 170:379–384, 1968.

[217] M. Kemmler-D. Koelle D. Bothner, T. Gaber and R. Kleiner. Improving the performance of superconducting microwave resonators in magnetic fields. *arXiv:1101.3185v1*, 1:4, 2011.

[218] T. McMillan, P. Taborek, and J. E. Rutledge. A low drift high resolution cryogenic null ellipsometer. *Rev. Sci. Instr.*, 75:5005, 2004.

[219] Material Properties. Stainless steel type 316ln austenitic stainless steel (uns s31653). Technical report, ATI Allegheny Ludlum, 2009.

[220] Anonymous. Metallic materials and elements for aerospace vehicle structures 2.7. Technical report, MIL-HDBK-5F, U.S. Department of Defense, 1987.

[221] Anonymous. Aluminium material data sheet en aw-1050a, en aw-al 99,5. Technical report, Aluminium-Verlag, Marketing & Kommunikation, Aachener Stra e 172, D-40223 Düsseldorf, 2009.

[222] K. D. Jayasuriya, A. M. Stewart, and S. J. Campbell. The specific heat capacity of ge varnish (200-400k). *J. Phys. E: Sci. Instrum.*, 15:885, 1982.

Part V.
Appendix

A Experimental

A.1. Performance and setup of the ARS closed cycle cryo-head

The complete cryo-system origins from ARS[1] and consists of a Giord McMahon cryo cooler, the corresponding compressor, and the He-pressure lines as shown in Fig.(A.1). Inset a) shows both, the compressor and the cryo-head connected with nearly 10m long He-lines. A closeup of the cryo-cooler is shown in inset b) which illustrates the two cooling stages where the outer 4K-cold finger is coupled via a He-bubble to the inner 2nd cooling stage which keeps the displacer assembly. The valve-motor, and the valve are situated at the top, both leading to vibrations in the upper, decoupled part of the cryo-cooler. The GMX20-B system is a modified version with the mentioned mechanical decoupling, to ensure the 2nd stage to be nearly vibrationless as demanded for cold-atom experiments.

Without major heat load from the environment[2], capable of cooling the 2nd stage down to slightly below 4K. This is either indicated by temperature sensors installed at the 2nd stage as well as the experience that He-gas in the decoupling bubble condenses.

A heat load performance measurement for the system was never performed, although a heating stripe at the 2nd stage is capable of up to 5mW, 500mW and 50W and therefore would allow to continuously ramp up the temperature due to heat load. Nevertheless the stepwise implementations into the experiment eectively increased the base temperature of the 2nd stage.

Typical temperatures measured with four dierent sensors, for a setup as it is finally realized in this thesis[3] the typical temperatures are given below:

- $T_{base} = 5.1400 \pm 0.002K$: temperature at the base rather close to the 2nd cooling stage
- $T_A = 5.1040 \pm 0.0005K$: measured at the lowest transport coil mounting
- $T_C = 5.0720 \pm 0.001K$: measured at the uppest transport coil mounting

---
[1] Advanced Research Systems, Inc., Macungie, PA 18062 - USA
[2] Such as electrical connections from 300K down to 4K and windows at the thermal radiation shield
[3] Realization: complete transport setup, 2 windows in the thermal shield, 40 wires down to the 2nd stage (without a current flowing)

A. Experimental

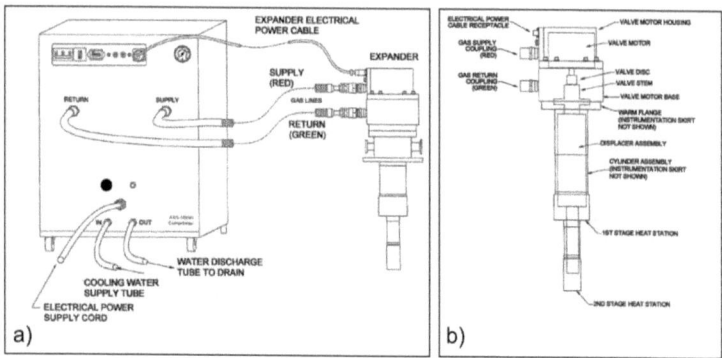

Figure A.1.: Inset a) shows the setup scheme for complete cryo-system and with compressor, cryo-cooler and He-pressure lines. Inset b) shows the inner life of the cryo-cooler with the 1st and 2nd stage and the mechanical valve and motor as used in Giord McMahon cryo-coolers. Both graphics are taken from the supplier manual for the GMX20 type machine.

- $T_{shield} = 54.26 \pm 0.01 K$: at the top of the uppest shield section, 10cm below 1st stage

The sensors provide a quite high stability, and are partially gauged down to temperatures of 4.5K, respectively 4K, or, as the sensor for T_A, gauged and calibrated by the supplier.
Fig.(A.2) shows the heat load performance as tested by the supplier, inset a). In addition the cool-down behavior of the complete system is shown, indicating that the total heat load is low enough to reach a base temperature of at least 5K. The big advantage of closed-cycle cryo system, is the comfortable handling that allows just by pressing a button, to cool the system down within less than 15 hours, as shown in inset b). The warm up, without heating the stages, just by self-heating takes less than 18 hours. As the cool-down is maintained while the heating stripes in the vertical transport section are up to 200° high, the oset temperature of the 1st and the 2nd stage are incredible low as $\Delta T_1 \approx 2K$ at the thermal shield and $\Delta T_2 \approx 0.1K$ at the 2nd cooling stage.

During the experimental cycle the temperature of course increases as shown in Fig.(A.3), whereas one time-tick lasts 20s. As shown in the inset, the periodic peaks do not correspond to the experimental period, but as the period last for 8 time-ticks, an aliasing occurred in the measured temperature. Nevertheless, it shows that the the maximum temperature increase (which is indeed monitored) is less than 150mK during the final experimental cycle transporting atoms into the cryogenic environment.

Notice: as the temperature-sensors indicate, the system is quite temperature-stable. Nevertheless, changing the pressure in the He belly, responsible for the mechanical decoupling, both via adjusting the distance of the two 4K cold fingers or the slight overpressure (1atm+40mbar), results in a temperature change even in the range of 10-100mK.

A.2. Windows at 50K

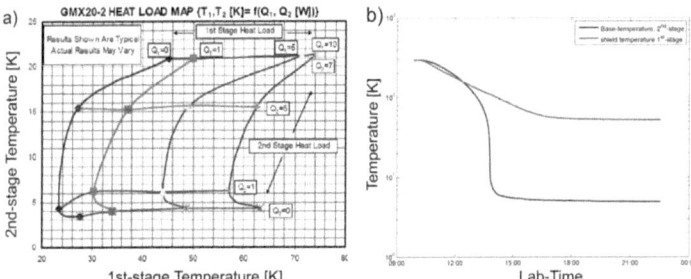

Figure A.2.: Heat load performance of the GMX20-B system, measured by the supplier, inset a) and cool-down characteristic for the fully setup experiment is shown in inset b) Cool-down is achieved in less than 15 hours, while self-heating up to room-temperature takes nearly 18 hours.

A.2. Windows at 50K

Non-birefringent windows

According to the design considerations shown in section (12), a proper mounting of the windows is crucial to prevent damage, and birefringent optical properties.
Following the *Haroche Group (ENS/LKB Paris)*, the windows are made of SF57 glass, which has a particularly low stress-induced birefringence [218]. Properly mounted, they give the total measured phase shift between orthogonal polarizations for a laser beam crossing the whole setup, which they point out to be below 0.3 rad. Up to now, the implemented windows were not tested regarding to the non-birefringence properties, which is still an open question, but not crucial[4] for the measurements presented in this thesis.

Broken windows

Windows broke during cool-down even if they were just very carefully tightened with little Teflon stripes as shown in Fig.(A.4), inset a). Furthermore the thermal contact is improved with a thin layer of Apiezon vacuum grease[5] but it turned out, that initially the inner diameter of the window-mountings was too small. Hence the diameter of the mounting was increased from 50.2mm to 50.6mm as the aluminum mounting turned out to shrink during cool-down. In addition the vacuum grease is now just applied at the circumference and not at the ring-shaped outer contact surface of the window[6], which decreases the mechanical stress appreciable. After this provisions, windows did not break anymore.

---

[4]For detailed arguments follow section A.8 below, which describes the method of detection in this experiment
[5]Apiezon Products, MI Materials Ltd., Hibernia Way, Traord Park, Manchester M32 0ZD, UK; www.apiezon.com
[6]The Apiezon vacuum-grease seems to get sti and tough, when cooled down.

A. Experimental

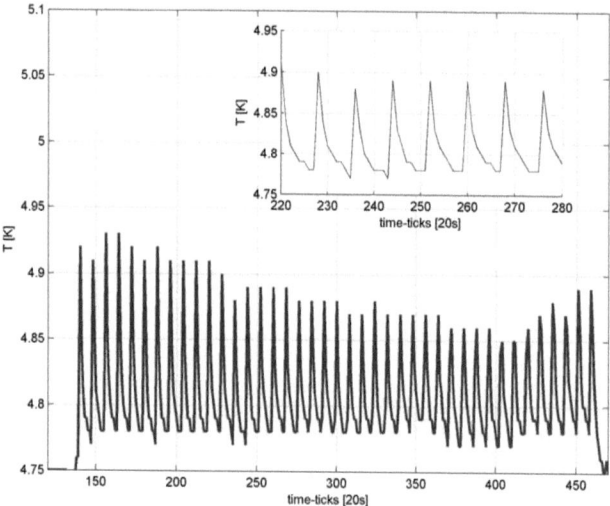

Figure A.3.: Base temperature during the experimental cycle. Even if one time tick lasts for 20s, the maximum temperature increase during the experimental cycle does not exceed 150mK. The duration of the experimental cycle eventually leads to an aliasing eect as the periodicity in the inset is given by 8 time-ticks. The increase obviously appears due to the current carrying transport coils in the cryo-system.

A.3. The challenge of SC coils

Figure A.4.: Windows and mountings at the thermal shield. Inset a) shows a previous way to tighten the windows, which lead to a bad thermal contact, and even in the beginning of the testings to broken windows. Inset b) shows a broken window, with a temperature sensor mounted, to measure the window-temperature. Inset c) and d) show how the tightening is now realized to avoid mechanical stress, while ensuring the window to be proper thermally anchored.

Thermal input from high-temperature windows

These issues resulted in the replacement of the Teflon-stripes to fasten the windows with a Teflon- and steel-ring as shown in inset c) and d). Inset b) shows how a temperature sensor which was mounted onto the window to check its temperature. As mounting of a temperature sensor directly on the window kills the latter, this provision was just applied on a broken one. Nevertheless before replacing the Teflon-stripes with the above mentioned rings, the temperature at the windows was as high as $\approx 140K$ even if the thermal shield itself was at $\approx 55K$. This huge temperature gradient originates on the one hand from the rather bad thermal properties of SF57[7], and the bad mounting of the window, in combination with the high heat input from the outside environment. According to Stefan-Boltzmann, this high temperature corresponds for the window to a radiated power of $\approx 2.2 mW/cm^2$ in contrast to $\approx 52 \mu W/cm^2$ if the window would be at $T_W = 55K$.

This seems to be a quite small heat load, nevertheless it correspond to a factor of 40 and becomes a major issue if the coils should get superconducting. This is now discussed in the following section.

A.3. The challenge of SC coils

In the following three issues are presented which made the fabrication of superconducting coils quite challenging. First, electric short occurred during winding, second and third: coils did not became superconducting at all, or did not sustain the desired current, often resulting in an irreversible destruction of the coil. The issues are presented in detail in the following.

A.3.1. Winding the coil

The first coils were wound using a vacuum grease (Apiezon) to maintain thermal contact between the windings. This seemed to be quite suitable, as the coil can easily be removed from the mounting, if it brakes during winding. To maintain even a better thermal contact, GeVarnish[8] was used in all later B-,C-, and D-type mountings as it turned out that GeVarnish

---
[7]For more details see Appendix D
[8]Thermal conductance of Ge IMI 7031 [02-33-001]: $63 mW/(m.K)$ at 4.2K, 10min air-drying if not thinned

A. Experimental

can easily be dissolved putting the mounting/coil into Isopropanol for several hours. During the winding process ethanol-thinned GeVarnish is applied between the windings, which is after the winding process baked out for 12 hours at $\approx 80°$.
For winding the first coils, it always turned out that there is an electric short between the coil and the Cu-mounting, measuring $R_{A-ground} + R_{B-ground} = R_{AB}$. This even occurred if the resistance was carefully checked every 300 windings. Sometimes the short even occurred after the baking-process, which is indicated by two reasons.
First, the mechanical stress which occurs do to the winding of 3000 coils can damage the wire insulation, if sharp edges at the coil-mountings are not carefully removed[9]. This might be enhanced as it is known that *GeVarnish* aects the *Formvar* insulation of the used sc-niobium matrix wires. This would explain why the electric short always occurred at inner windings, as indicated by $R_{A-ground} \ll R_{B-ground}$.
Therefore an electrically insulating layer was introduced between the coil and the Cu-mounting consisting of a very thin mylar foil which is glued in between using GeVarnish.

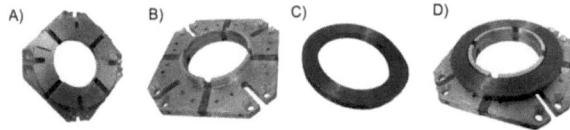

Figure A.5.: Inset A) shows a coil wounded on a B-type mounting, whereas inset B-D shows the modified D-type mounting and a self-sustaining transport coil glued onto the sliced mounting, glued together with a thermal conductive Epoxy

A.3.2. Dissipation at the solder contact

The critical parameter for the sc-coils is the current which it can withstand. In principle this is just limited by the critical current of the Nb-Cu matrix wire defined by the Nb-filaments. Nevertheless the sc-wire is somewhere connected to a normal conducting wire, and the solder contact can dissipate heat, which must be lead away.
It turned out that, when characterized, the coil resistance (R_{cu}) increased with current (indicating dissipation in the normal conducting part), but if a critical point is excessed a runaway behavior of the resistance is observed. This turned out to be a irreversible process as the maximum current in later tries was always smaller or even much smaller than before.
A simple model support this assumption using measured data from as shown in Fig.(A.6), inset a).

Dissipated heat through the solder contact from the coil current I_{B02}, continuously heats the normal conducting (nc) wire. Starting with 5K equilibrium temperature of the base-plate thermal bath, just a small part dl contributes to the ohmic heating in the nc-wire. As the wire is anchored at a distance Lw from the solder contact the dissipated heat must be transported through the wire. Accounting for the increasing temperature at the solder contact and along the wire, together with the increase in ohmic resistance along the nc-path, the temperature in

---
[9]Intrinsically the slicing of the mounting introduces edges and corners

A.3. The challenge of SC coils

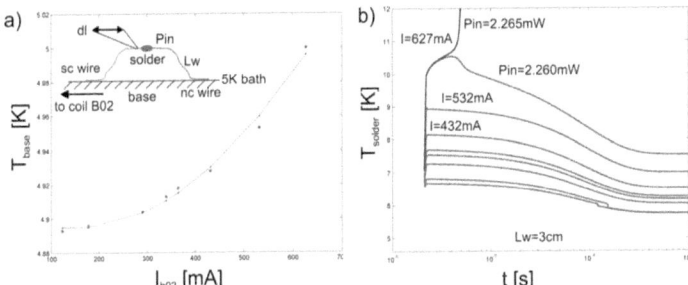

Figure A.6.: Inset a) shows the measured base temperature increase with coil current ( ) a numerical fit, which shows a quadratic behavior. This indicates that dissipated heat from the normal-conducting (nc) Cu-wires affects the base temperature. This inset also shows the physical model to describe the modeled avalanche effect. This effect occurs if the length Lw between the solder contact and the anchoring on the base-plate is too long (3cm). It is completely sensitive to a certain threshold of input power, such as the dissipated power at the solder-contact, inset b) where an estimation of the temperature at the solder-contact is shown for different coil-currents.

the solder-contact can recursively be calculated. If the initial heat input through the solder contact which is set $P_{in} = 2260 \mu W = const.$, is just slightly increased, the heat transport through the wire can not be done fast enough and the solder contact heats above the critical temperature for Nb. At this point an avalanche starts, as more and more wire segments dl of the sc-wire plays a role for the heat input $P_{total} = P_{in} + P_{nc} + P_{sc}$, which dramatically increases the temperature along the sc-wire. Inset b) of Fig. A.6 shows for $Lw = 3cm$ (length between solder contact and anchoring on base-plate) the avalanche effect which is completely sensitive to the dissipation in the solder contact.
As the non-anchored wire length is that crucial, the maximum coil current could indeed be increased to $1.35A - 1.5A$ for dozens of seconds, if the non-anchored wire length Lw is reduced to $Lw = 1mm$. For those simulations the avalanche would even not start before $I_{B02} = 1.5A$ is reached.

It turned out, that the optimal solder contacts for the 4K section are Cu-tubes which are filled with solder, plugging in the wire ends from both sides. Even because of the above mentioned avalanche effect, the Sub-D connector version did not work at all, as well as just twisting the pairs instead of using a Cu-tube was not reproducible successful. BeO heat sink chips[10] which provide a good thermal contact right at the solder connection turned out to be too tiny to be flexible and often used, as well as the surface behavior does not allow for proper soldering at all.

---

[10] BeO heat sink chips from Cryophyiscs GmbH Dolivostra e, D-64293-Darmstadt

A. Experimental

A.3.3. Radiation input from the windows

After single coils were repeatedly driven up to 1.5A, an issue occurred building in all four transport coils simultaneously. It took quite a long time to figure out, that the coils even reach different maximum currents at different slot positions which lead to the educated guess that there might be some additional heat input. The windows are known to be far above the shield temperature, nevertheless introducing just a rather small radiation power in the order of $\approx 2.2 mW/cm^2$ at the approximated temperature of 140K. The following estimation shows that even this small power density radiated onto the coils can cause the punctual wire temperature to exceed even the critical temperature, leading to a vanishing superconductivity in the coil, if the coil, or even some outer windings, are badly thermally anchored and the window is too hot.

Assuming a thermal conductivity for the Cu-matrix of the superconducting coil-wire of $\lambda = 50 W/mK$, a wire diameter of $d_W = 112 \mu m$, and a length $l = 5cm$, the equilibrium temperature of the outer layer of coil-windings for a considered volume $d_W \cdot l$ could be estimated.

$$P_{abs} = A \times p_{rad} \cdot d_W \cdot l \tag{A.1}$$

Caused by $p_{rad} = 2.2 mW/cm^2$ from a window at 140K, the absorbed radiation power P_{abs} in a wire volume ($d_W \cdot l$) next to the window, leads effectively to an equilibrium temperature in this wire part as just $P_{away} = \frac{d_W^2}{4L} \times \lambda_{Cu} \Delta T$ is transported towards the $T = 5K$ bath along ≈ 20 last windings and the 30cm long connection towards the solder-contact, thus assuming a length of ($L \approx 3000mm$) for transporting P_{away} to the thermal base[11]. With Eq.(A.1), this leads to the estimation of the maximum wire temperature as described by

$$C_V \cdot V_{wire} T_{wire} = \int_0^t (P_{abs} - P_{away}) dt \tag{A.2}$$

assuming the heat capacitance C_V of the wire, and the heat conductivity λ_{Cu} to be not temperature dependent in this case. In addition thermal conductance through the GeVarnish is neglected. Furthermore this results in

$$\frac{\partial}{\partial t} T_{wire} = \frac{1}{C_V (d_W \cdot l)} \times \left(P_{abs} - \frac{d_W^2}{4L} \lambda_{Cu}(T - T_0) \right) \tag{A.3}$$

Finally this leads to the simple differential equation

$$\frac{\partial}{\partial t} T = A - BT \tag{A.4}$$

with the constants A and B as

$$A = p_{rad} \frac{4 A}{C_V \cdot d_W} - \frac{T_0}{C_V l \cdot L} \tag{A.5}$$

---

[11] Worst case for a winding in the uppest layer, which is not the last winding of the coil, and therefore even ≈ 20 windings are left until the 3000 turn are over.

A.4. Transport stability

$$B = \frac{}{C_V l L} \qquad (A.6)$$

derived from Eq.(A.3). Assuming $T_0 = 5K$, the solution of A.4 is as simple as

$$T(t) = T_0 + T_\infty (1 - e^{-\frac{t}{\tau}}) \qquad (A.7)$$

and leads for $t \to \infty$ to the temperature T of the small, 5cm long wire part l next to the window in equilibrium, which is derived as $T_{equil} = T_{inf} = A/B - T_0 \approx 180 \times \epsilon_A [K]$, independently on the specific heat capacitance C_V and depends on the parameter as

$$T_{equil} \propto \epsilon_A \times p_{rad} \frac{L^2}{d\ W} \qquad (A.8)$$

Regarding to a window which is at 55K instead of 140K with a radiation power density p_{rad} of just $\approx 50\mu W/cm^2$ compared to $\approx 2.2 mW/cm^2$, the equilibrium temperature would be much smaller and just slightly above the bath temperature in the order of $T_{equil} = T_\infty \approx 4.7 \times \epsilon_A [K]$, where ϵ_A results from the emissivity of the window and the coil wire and can be assumed to reach 50% in the worst case.

This temperature increase of the coil occurring at high window temperatures is indeed a problem and can easily occur if several criteria match: a) coil at a modified mounting without cap is used, which lead to a bad thermal contact for the outer windings; b) the coil is in one of the upper two slots in front of the windows and c) it is a self-sustaining coil badly glued onto the mounting.

Some of the coils were even not superconducting at all, at the uppest position (T_{coil} 10K or above), whereas they maintained 1.5A at the lowest coil slot, even on a modified mounting without cap!

In principle this issue was solved by several provisions:

- replace window blindings with high purity Al

- improve thermal contact between window-mountings shield

- improve tightening of the window into the mounting to increase thermalization

- wind Al tape around the coil to prevent direct radiation from the window on the uppest windings (which are just by several layers of GeVarnish contacted to the Cu)

- introduce 4K-Cu-blindings in front of the windows to prevent thermal radiation

In fact those provisions than resulted in working coils, whereby the heat input of the windows onto the coils was not the only reason why some coils did not work, but it played a certain role in the challenge for superconducting coils.

A. Experimental

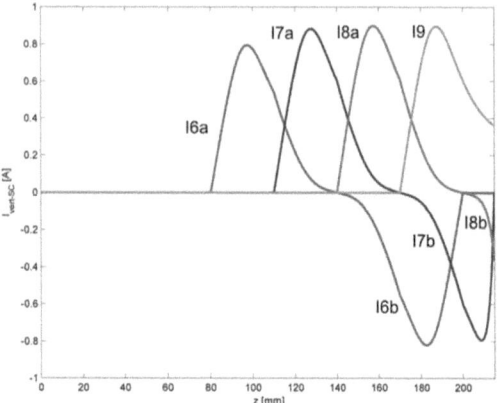

Figure A.7.: Superconducting vertical transport currents as they were the starting point for the vertical transport scheme. The positive and the negative half-wave of the different currents both belong to the current running through the same coil, but maintained from different current sources.

A.4. Transport stability

Based on the originally calculated sequence of vertical transport currents as shown in Fig.(A.7), the magnetic field gradient in vertical direction along the transport axis can be calculated.

As for every vertical position $z = f(t)$, the coils participate with the corresponding current I(t) from the coil position z', the minimum can be found solving

$$B_{min} = \sum_{i=6}^{9} B_i \left(I_i(z_{min}, z') \right) \qquad (A.9)$$

For every single time step a deviation from the supposed minimum can be found if either the transport currents or the coil geometry are changed, resulting in $\Delta z = z - z_{min}$ as shown for four different calculations in Fig.(A.8). Based on the supposed geometry shown in Tab.(10.2), and the real coil geometries shown in Tab.(A.1), the difference in trapping gradients can be calculated.

The overview is organized as follows. The quasi-linear behavior in the upper graph of inset a) shows the comparison of z and z_{min} on the x- and the y-axis which seems to be almost linear. Nevertheless a closer look shows that there is a small deviation which leads to a small oscillation of the atom-cloud around the supposed minimum as shown in the graph below. The third graph gives the vertical trapping gradient ∇B_z vs. the vertical trapping position which is supposed to be constant, but of course as the position varies also the gradient is different from the proposed 130G/cm.

A.4. Transport stability

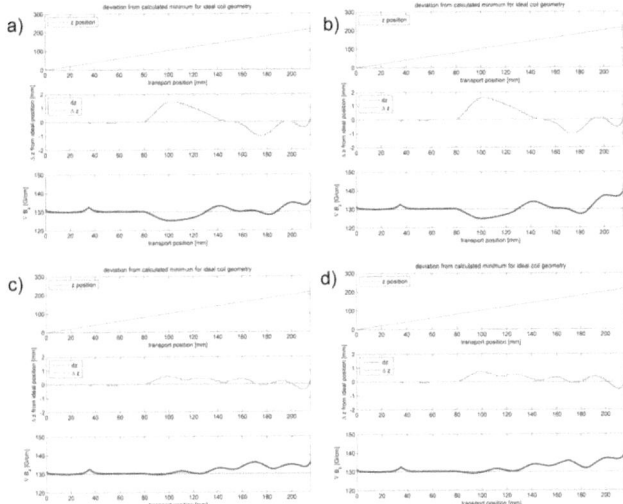

Figure A.8.: The figure depicts the vertical transport parameter, the calculated deviation from the supposed minimum and the vertical trapping gradient for the original transport currents and ideal coil geometries, inset a) and real coil geometries, inset b). For an increased current $I_6 = 10\%$ I_{6-0}, both lower insets gives the results for ideal, inset c) and real, inset d) geometries.

As indicated in inset a), the current I_6 was in the experiment decreased as the distance between the last nc-coil and the first sc-coils was assumed to deviate a bit from the desired distance $d = 30mm$. Nevertheless for the proof of stability the calculation of the deviation in the gradient and the position was done for equally spaced coil-distances and the preset current I_6, therefore resulting in the described deviation from 0-deviation line in inset a). Inset b) therefore compares the same currents with the real coil geometry, even leading to vertical gradients not much dierent.
Trying to shape the transport current I_6 therefore leads to a better result in the vertical trapping position and even to a more constant vertical gradient[12]. Inset c) and d) therefore show the results for ideal c) and real d) coil geometries and a vertical transport current $I_6 = (I_{6a} + I_{6b}) \times 1.1$ increased by 10%.

The interesting result therefore is, that this shaping of the transport currents did not lead to a significant change in overall transport e ciency. Nevertheless the final temperature of the cloud after transport was not measured.
In addition more shaped currents were applied. The complete variations are listed in

---
[12] For the robustness testing, just the vertical transport gradient is considered and calculated

A. Experimental

coil	V_6	V_7	V_8	V_9
$R_i[mm]$	24	24	24	24
$R_o[mm]$	35.1	35.1	36	34.4
$h[mm]$	5.3	5	4.9	5.7
N_{axial}	38	37	35	41
N_{radial}	79	82	86	74

Table A.1.: Real coil geometries for coils participating in the superconducting vertical transport: R_i, R_o are measured inner and outer radius of the coil after fabrication, h the real hight, N_{axial} and N_{radial} the number of windings in the corresponding directions, that most likely corresponds to the dimension of the coil body with the assumed wire diameter

Tab.(A.2).

Trial	V_{6a}	V_{6b}	V_{7b}	V_{8a}
A	+10%	–	–	–
B	+20%	–	–	–
C	+15%	+15%	–	–
D	+15%	+5%	–	–
E	+15%	+10%	–	–
F	+10%	+10%	–	–
G	+15%	–	+5%	-5%
H	+15%	–	–	-10%

Table A.2.: Dierent trials to shape the superconducting transport current scheme, just by multiplying the partial sequence from at least one power supply with a constant factor

Dierent voltage supplies were tested which are used at the constant current sources to provide the transport currents. As this current sources even act as a low noise constant current source, it obviously made no dierence in performance if car-batteries or rather dirty lab-voltage supplies were used.

Whatever was changed, the power supply (noisy or not) of the current source[13], or an deviation from the calculated transport scheme for the superconducting coils was introduced, the final number of transported atoms stayed rather constant within the experimental drift of ± 10%. Even lifetime measurements performed after the dierent trials showed no dierence, respectively increase or decrease in performance, turning out that the transport is in principle rather stable to markable deviations in the transport scheme.

A.5. Lifetime issues in the cryostat

In the beginning, after successful transport was maintained, the lifetime in the final superconducting quadrupole-trap was even bad, roughly 3-5s. This rather bad value was surprising, as in the cryogenic environment the pressure is supposed to be rather low. Nevertheless, an

---

[13] The current source itself provides rather low noise, independent of the voltage which is used to supply it.

A.5. Lifetime issues in the cryostat

additional lifetime-measurement in the corner of the transport section indicated a lifetime of $\approx 13s$ which lead to the assumption that three possible issues could be responsible for the bad lifetime in the cryo.

A) **Noise induced trap losses** in the quadrupole trap which occurred due to the heating of the atoms as the trap shape starts to oscillate. This was indicated as a noise-measurement of the system figured out that the cryo-cooler itself increases the background noise from $-145 dBV/\sqrt{Hz}$ up to $-110 dBV/\sqrt{Hz}$ in the range up to 20kHz just by switching on the cryo, with a huge peak in the region of 300Hz.

B) **Straylight** could lead to light induced collisions in the trap introducing huge trap losses which manifest in a short lifetime.

C) **Bad vacuum pressure** could be the reason as the cryogenic environment is not that good decoupled from the outer vacuum chamber. There are two possible connections, first the differential pumping stage at the bottom which connects to the transport section, to both, the inner cryogenic part and the outer vacuum chamber, and second several wholes at the top of the thermal shield.

Indicated from a 13s lifetime in the transport corner, and a vacuum pressure in the outer chamber measured to be $\approx 2 \times 10^{-9} mbar$, three different improvements were investigated finally leading to the lifetime presented in this thesis, far beyond 100s.

A.5.1. Improved differential pumping stage

As the cross section, connecting the transport with the outer cryostat vacuum chamber, of the differential pumping stage is even larger than the connection to the inner cold chamber, the assumption was that even the bad vacuum influences the inner chamber. Therefore three Kapton-rings were implemented, closing the space between differential pumping tube and transport tube nearly completely. In addition all holes on the top thermal shield supporting the pumping of the inner cryo-section with the turbo-pump were closed, except 2 of 24.

A.5.2. 4K cryogenic pump using charcoal

Charcoal is knwon to be used with cryogenic pumps as it increases the pumping surface due to its porous body. Thermalized to low temperatures, it acts as a huge cryogenic physi- and chemisorptive surface capable of decreasing the pressure in those systems dramatically. As the 4K surface at the experimental stage was rather small, compared to the surface of the thermal shield even at 50K, this was figured out to improve things dramatically. Three containers filled with charcoal were built in, filled with charcoal pellets. However in connection with the improved differential pumping stage and the closed holes to the outer chamber, this lead to a smaller pressure in the outer cryo-chamber of $8 \times 10^{-10} mbar$. In a first approach, this directly lead to an increase in lifetime to about 12-15s. Nevertheless it could not be figured out if this also lead to the finally measured lifetime of more than 100s.

A.5.3. Pulsed cryostat

A test was performed to check whether the continuous running cryostat induced trap losses due to its emitted noise-spectrum, in a setup when the charcoal was not yet implemented. Starting from a constant final atom number after transport of $\approx 4 \times 10^7$, the cycle was increased by 7s lasting in 7s cryostat running during MOT, than switching the cryostat off via a TTL triggered signal and keeping it off during transport with a recovering time

A. Experimental

afterwards, in total for 9s. This applied scheme resulted after 50 experimental cycles in an increase of the radiation shield temperature $T_{shield} = 56K \rightarrow 64K$ and a remarkable drop down in atom number N$\approx 4 \times 10^7 \rightarrow 1 \times 10^7$, thus indicating, that the radiation shield significantly participates in the achieved pressure in the cryo causing trapp losses through background collisions.

A.6. Eddy current issues and oscillating magnetic fields

When the atoms are imaged in-situ in the superconducting quadrupole trap, they are of course detuned due to the Zeeman-shift (see section 5.3.1). Trapped in $|F = 2, m_F = 2\rangle$ the atoms are detuned $1.4 MHz/G$ and as the trap is located in a wide space-region, most atoms see $B \neq 0$ which means that they are imaged o-resonance.

As all coils, respectively the quadrupole coils are switched o after a certain hold-time in the trap, the quenching circuits (Appendix B) in the 5A-switches ensure the superconducting currents to be ramped down quite fast, even within 1ms. First it was not clear, that as shown in Fig.(14.1), the absorption signal of the imaged atoms vanishes and comes back after a certain TOF. Nevertheless, the detuning scan from Fig.(14.2) indicates, that the reason is a strong magnetic field which detunes the atoms. After almost $TOF = 4ms$, the magnetic field has even canceled and a maximum peak of atom number can be recorded. Nevertheless both, the TOF-scan in Fig.(14.1) and the 3D detuning-TOF Fig.(14.2) show signal-oscillations, with time and detuning.

Hence this is quite interesting as the fast switch o, of the magnetic trapping fields, obviously induces a magnetic field at the position of coil V_7, and therefore induces a voltage which can be measured Fig.(A.9), inset a). This figure shows, and even the closeup inset b), that there is a strong evidence for an oscillating magnetic field (even slightly phase shifted in V_7) that is at the trap center responsible for the oscillating behavior in a series of TOF-images, inset c), also possibly resulting from not simultaneously quenched trapping currents.

Useless to mention that even a superposition of eddy-current loops which are induced at several distinct places in the cryogenic setup can not create an oscillating field. In addition Fig.(A.9), inset a) shows that the oscillation starts even milliseconds after the quenching circuit has enduringly canceled the coil current.
Nevertheless one possibility remains, based on simple electric circuit dynamics. Assuming an parallel LRC-circuit consistent of the coil inductance L_{coil} and the capacitance C_{coil} maintained by 3000 windings and insulated by a $20\mu m$ thick *Formvar* insulation, results in an resonance frequency of the circuit. This resonance frequency therefore could be related to the oscillation period of roughly 0.9ms measured in the setup as shown in Fig.(A.9). Therefore the resonance frequency

$$f_0 = \frac{1}{2} \frac{1}{\sqrt{L_{coil} C_{coil}}} \tag{A.10}$$

with $L \approx 600mH$ and the capacitance according to Lecher-lines, results in $f_0 \approx 800Hz$ for the chosen setup with

$$C = {}_0 \frac{l}{arccosh(1 + \frac{A}{2R})} \tag{A.11}$$

A.6. Eddy current issues and oscillating magnetic fields

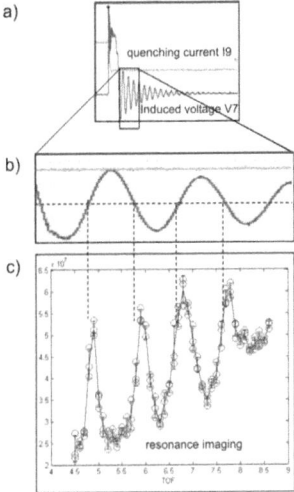

Figure A.9: Inset a) shows the measured induced voltage at coil V_7 and the corresponding quenching current I_9. This inset depicts that the ramp-down of the coil current within 1ms, even a corresponding fast oscillation of the induction voltage still can be measured at coil V_7. Nevertheless, the damped induction voltages as depicted in b) seems not to be related to any electronic artifact in the 5A-switch as it even builds up milliseconds after canceling the trapping currents. Inset b) even shows a closeup of this induced oscillation voltage, and inset c) the corresponding oscillation of atom-number in a TOF-measurement at zero detuning from resonance.

which defines the capacitance between two parallel wires with length l, radius R, and mean distance d, with $\Delta = d - 2R$, resulting in $C_{coil} \approx 67nF$ for a superconducting coil. In addition the resistance of $FormVar$[14] between the windings could even be in the order of several $1k$ or below.

Assuming eddy-currents, they would enter as additional resistive impedance loops in parallel to the LC, respectively LRC circuit, leading to a damping-time constant of $\approx 30ms$ as measured in Fig.(A.9) due to

$$= \frac{2L_{coil}}{R_{total}} \tag{A.12}$$

Consequence of these considerations are that 1) the eddy-currents themselves can not be responsible for the observed oscillations in the TOF-scans where atoms are obviously detuned due to revivals of magnetic fields, and 2) that the self-resonance of the superconducting vertical transport coils with N=3000 (acting as an LRC-circuit) in the order of several 100Hz could explain the oscillation period as well as the characteristic time in which the oscillations vanish if low inter-coil resistance of the $Formvar$-insulation and damping due to eddy-currents is considered. In addition a non-simultaneous quench of the trapping-current could be avoided if a serial-circuit power splitter is introduced to drive a serial current through all coils participating in the quadrupole- or QUIC-trap.

Notice that even a comparison of the TOF-scans from Fig.(A.9), Fig.(14.2) and Fig.(14.1) shows different behavior, as they were recorded after the internal environment [15] was changed,

---

[14] FormVar is known to be fragile to GeVarnish which is used as a glue and which can even dissolve the insulation of the superconducting wires: $\rho_{FormVar} \approx 10^{14} - 10^{16} \cdot cm$ at $T = 300$

[15] Building in the Ioe-coil and mounting for example, changes the behavior of the oscillations completely as

A. Experimental

which obviously results in more or less dierent eddy-currents-behavior.

A.7. Ingredients to build up the 4K environment

The following summary presents dierent cooking ingredients were implemented during the setup of the 4K cryogenic system. Material parameter can be found in Appendix (D) for further details.

GeVarnish (IMI-7031) is an epoxy and glue for mechanical fixing and insulation at cryogenic temperatures. It is used for winding the coils even to maintain good thermal conductance between the windings, and for fixing the superconducting wires to bobbins for thermal anchoring. It dissolves in ethanol and is curable at room temperature.

Apiezon N is a silicone-free vacuum grease for high vacuum that ensures cleanliness and low out-gassing without contaminating vacuum systems. It has a high creep-resistance and a quite high thermal conductivity at 4K. Therefore it is used to contact the windows in the window-mountings and additional the thermal shield connections. It is a good choice to thermally anchor temperature sensors.

Silver foil is mechanical very soft and an outstanding good thermal conductor. It is used in between the thermal shield parts as Aluminum is quite soft and may warp if tightened too strong, resulting in a bad metal-metal thermal passage.

High purity Aluminum is used for all thermal shield parts, including the blindings, the dierential pumping stage and the window-mountings. It is better machinable than high purity Cu and provides a higher electrical resistance which is well suited for less eddy-currents. The thermal conductivity is at 50K high enough but even lower than in Cu.

High purity Copper provides highest thermal conductivity. It is used for the superconducting coil mountings, for the coil-cage, and for the experimental stage at 4K.

Stycast is a two component epoxy. The 2850FT/11 type provides hardness, high thermal conductivity and electric resistivity. It is used to glue the sliced coil-mountings together.

Aluminium tape is used to fix wires. It can easily be removed and thermally connects things together. It is also wound around coils and coil-mountings next to the windows mounted at the radiation shield to prevent direct radiation from the window onto the coils.

Solder for connecting the normal- and the superconducting wires should even provide an extremely low resistance. Therefore a solder of type HF32, S-Sn60Pb39Cu1 is used for all solder-contacts in the cryogenic environment.

Charcoal is used to improve the cryogenic pumping in the 4K-area. In this setup 3mm pellets which are already steam-activated are under use.

Teflon tape is used to tighten flying around wires and increase thermal contact as everything is wrapped around the cooling stages. In addition it is easy removable which makes it a first choice-tool if something should/could not be screwed or glued.

Kapton tape and foils are used as thermally conductive, and electrical resistive materials to fix wires to anchor-bases. In addition it allows by definition, to insulate metal-parts without aecting thermal conductivity much.

---

eddy-currents behave dierent

A.8. Method of detection: Absorption Imaging of cold atoms

The method of choice to detect the atoms is resonant absorption imaging. In contrast to fluorescence imaging where just a fraction of emitted light is collected in the imaging system, the complete relative intensity of light absorbed by the trapped atoms, is measured. Therefore this method is a destructive one. The following briefly gives an overview of the most important things to consider when using absorption imaging, as this method is well known and a standard technique in cold atom physics [188].

Absorption Imaging

The attenuation of a resonant laser-beam is given by the Lambert-Beer-law by

$$dI = -I \sigma_{abs}(\delta, I) n(r) dz \tag{A.13}$$

with the atom density $n(r)$ and the laser-beam intensity I for a beam propagating in z-direction. Hence, the absorption coefficient is given by

$$\sigma_{abs}(\delta, I) = \sigma_{abs}^0 \frac{\alpha}{1 + \frac{I}{I_s} + 4\frac{\delta^2}{\Gamma^2}} \tag{A.14}$$

were $\sigma_{abs}(\delta, I)$ is for $\sigma^\pm$ polarized light using the resonant absorption coefficient for a two-level system $\sigma_{abs}^0 = 2\lambda^2/2\pi$ with the imaging wavelength λ. This absorption cross section is even valid for an transition with certain sub-levels such as the m_F-manifold of the hyperfine levels $F = 2 \to F' = 3$ due to the Zeeman-shift, which is used as the imaging transition in this thesis.

Therefore the atoms should be imaged with circular polarized light regarding to the imaging axis. In addition this demands a small magnetic guiding field during imaging. In this case the coefficient would be $\alpha = 1$, where for the case that the atoms are equally distributed over the sub-levels, an average over the Clebsch-Gordan coefficient would exhibit $\alpha = 7/15$ [106].

In this work, neither a magnetic guiding field is used during TOF-imaging, or in-situ imaging[16], nor the imaging light is right-polarized. It is rather a superposition of circular polarized light, as linear polarized light is in-coupled into the fiber, and directly sent through the vacuum-chamber after the out-coupler, see Fig.(A.10). Therefore atom numbers measured in this thesis are in the case of TOF-imaging an underestimation due to the following consideration:

Following the Labert-Beer law Eq.(A.13), the 2D-transmission reads as

$$T(x, y) = \frac{I_f(x, y)}{I_0(x, y)} \tag{A.15}$$

and further

$$\frac{I_f(x, y)}{I_0(x, y)} = e^{-\sigma_{abs} \int_{-\infty}^{+\infty} n(r) dz} \tag{A.16}$$

---

[16]This would obviously make no sense as an additional guiding field would cause changes to the trapping potential. If the trap would be an Ioe-like trap, in-situ-imaging intrinsically performs with a guiding field (depending of the field strength)

A. Experimental

with the final intensity I_f, and $\int_{-\infty}^{+\infty} n(r)dz \to \bar{n}(x,y)$, the experimentally measurable column density, the optical density can be introduced,

$$\mathfrak{D} = \sigma_{abs}\, \bar{n}(x,y) \tag{A.17}$$

proportional to the absorption cross section σ_{abs} and the column density $\bar{n}(x,y)$. As the total atom number can be calculated from the integral of the column density $N = \int\int \bar{n}(x,y)\,dx\,dy$ the imaged atoms reads as

$$N = \int\int \bar{n}(x,y)\,\sigma_{abs}(\sigma, I)\,dx\,dy \tag{A.18}$$

Using Eq.(A.14), and Eq.(A.16) and the restriction that $I/I_S \ll 1$ for the saturation intensity I_S, the atom number simplifies to

$$N \approx \frac{1}{\sigma_{abs}^0}[1 + 4\frac{\sigma}{2}^2]\int\int \ln\left(\frac{I_0(x,y)}{I_f(x,y)}\right)dxdy \tag{A.19}$$

This now exhibits that for non-circular polarized light, and no magnetic guiding field applied, the atom number is always an under-estimation as $\sigma < 1$ would be valid. To exclude over-estimation of the trapped atoms, σ is set to $\sigma = 1$ in the algorithms implemented in the experimental data-acquisition.

For practical reasons the absorption images are calculated following a background correction. As the imaging cycle contains three exposure pulses with I_{cloud} resulting in a picture with the shadow of the atoms as they have absorbed partially the resonant imaging beam $\sigma = 0$, I_{beam} without atoms and a background picture I_{back}, the number of atoms is calculated summing up the pixels as

$$N > \frac{A}{\sigma_{abs}^0}\sum_{x,y}\frac{I_{cloud} - I_{back}}{I_{beam} - I_{back}} \tag{A.20}$$

where the factor $A = \frac{pixelsize}{g_m} = [m^2]$ contains the camera pixel size and the magnification g_m, and the absorption cross section $\sigma_{abs}^0 = 2.9 \times 10 - 13 m^2$.
Due to technical limitations the duration between the imaging pulses is $\approx 200ms$, even a long time, where the setup can jiggle and thus may introduce interference patterns in the pictures.

Imaging setup realized in the experiment

The two independent imaging system are shown in Fig.(A.10). They are realized in a very simple setup. Even the lower imaging system provides circular polarized light, the atom numbers in this thesis are an under-estimation as both, the number is calculated with $\sigma = 1$, even if no guiding field was applied during imaging, and second the imaging beam shares one quarter-wave-plate with the optical pumping setup. As the optical pumping was optimized for transport into the cryo, images taken after optimization for transport do not provide the right polarization. Inset a) shows the principle of the absorption imaging, with magnification achieved via an objective. The magnifications realized in the setup are shown in inset b), realized with commercial camera objectives.

A.8. Method of detection: Absorption Imaging of cold atoms

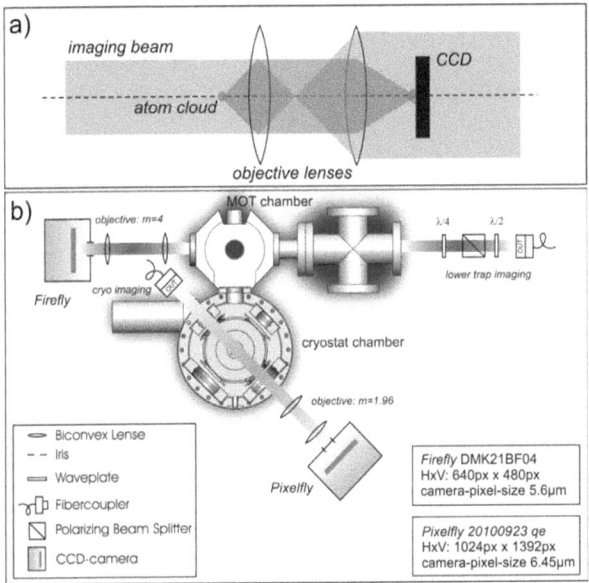

Figure A.10.: Inset a) shows the principle of absorption imaging. The complete experimental imaging setup is depicted in inset b). It shows the two independent setups for the lower magnetic chamber and the cryostat chamber were imaging without circular polarized light is implemented.

Temperature Measurement of an atom cloud

The temperature of a cloud of trapped atoms is determined from the thermal expansion of the cloud during free fall. Therefore the trapping fields are switched o fast, and the atomic cloud expands during free fall in gravity. The expansion of the cloud during the center of mass motion is purely related to the thermal velocity distribution if a non-interacting gas is assumed, where the velocity and hence the kinetic energy origins from $k_B T$ only. After a certain time-of-flight (TOF), absorption imaging is applied. Considering a macroscopic trap with trap-frequencies almost equally and hence similar velocity distribution in all three directions ($\sigma_{(v_x, v_y, v_z)} = \sigma_v$), results in an isotropic expansion with

$$\sigma(t) = \sqrt{\sigma_0^2 + (\sigma_v t)^2} \tag{A.21}$$

and hence the full-width half-maximum diameter of the cloud diameter $\sigma(t)$ after $t = TOF$ assuming a Gaussian velocity distribution with FWHM σ_v. Therefore the temperature can be found from

217

A. *Experimental*

$$\sigma_v^2 = \frac{k_B T}{m_{Rb}} \quad \text{(A.22)}$$

which finally leads to

$$T = \frac{m_{Rb}}{k_B t^2}\left(\sigma^2(t) - \sigma_0^2\right) \quad \text{(A.23)}$$

B Electronics

Switch-boxes for magnetic transport

Beside the switch-boxes, which contains the main electronic necessary for the room-temperature transport, the temperature-control box measures the coil temperatures. The latter, is in fact a micro-controller board which processes the measured temperatures and sends an error-signal to all switch-boxes in case of an error. Therefore this box should not be considered in detail. The hardware of the switch boxes consists of the main-board with golden rails and alternatively implemented MOSFET-switches, H-bridges, and short-circuits as shown in Fig.(B.2). These short-circuits are used to already extract a current from the main current-power-supplies while the coils are still switched o.

The electronic contains a de-multiplexer board which allows to multiply control one power supply during dierent phases of the transport, a demux-control board which en-/disables the electronics, and a current monitor.

Fig.(B.1) shows the inner life of a switch-box, even without H-bridges. Inset a) gives an overview of the box. Following parts are ascending labeled:

1. De-multiplexer-board with 4-digital input channels. This in principle allows 16 outputs to be driven with one control-signal since the channels are encoded binary ($2^4 = 16$). At least not more than 3 ouput-channels are used in one switch-box.

2. Current monitor which measures the output current.

3. Demux-control board which measures the Cu-rail temperatures and receives an error signal from the temperature control box if a single external temperature sensor exceeds the limit. If an error signal is received, the switch-box disables the corresponding power-supply and waits to be reset.

4. Short-circuit bridge as it is used to extract current from the power supply just before the MOSFET switches opens, as this allows a fast slope ramping up the current.

5. Two batteries of MOSFET switches for two dierent coils. As the MOSFETs withstand 30A, four of them are applied in parallel.

B. Electronics

6. Digital trigger as inputs for the de-multiplexer board, which enables the encoded output.

7. Output-lines for two coils.

8. Monitor output for the current running through the Cu-rails.

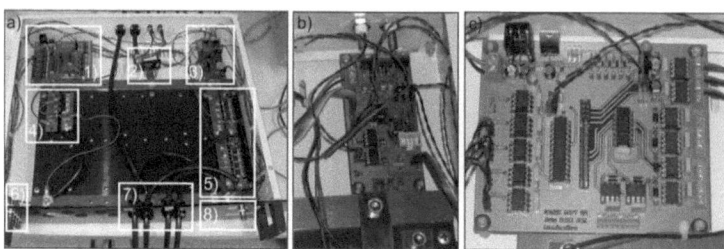

Figure B.1.: Inset a) gives an overview of the inner life of the switch-box. Beside the Cu-rails, two electronic boards are implemented. The de-multiplexer board which allows to enable one power-supply for dierent coils in dierent phases of the experimental cycle, and the demux-control board, which processes error signals from dierent temperature sensors, internally and externally measured. Inset b) shows a closeup of the latter, while inset c) shows the closeup of the de-multiplexer circuit.

Quenching circuit for highly inductive coils

To switch the highly inductive superconducting coils fast, and much more important, to protect the coils from induction voltages, a switch capable of up to 5A is used with an implemented quenching circuit. The switch-layout origins from the electronic workshop from the physics faculty at the University of Heidelberg, and was adapted. First, a 1W resistance with $R = 1$ was introduced in the quenching line, indicated with 1), see Fig.(B.5) to shorten the characteristic time. It was found that originally the switch demanded 5ms to linearly ramp down the current through the superconducting coils. As the intrinsic resistance of the cold coil is ≈ 0.5, it was quite obviously to implement a serial resistance after the diode $D1$ and the Darlington-transistor, and indeed the circuit showed a shortened switch-o resulting in $_{off} = 1ms$[1]. In addition a fast melting fuse with $I_c = 0.5A$ was implemented in serial to the power-MOSFET, to protect the sc-coil to be charged with more than the maximum current[2] of 1.5A. Therefore it is ensured that in any case the connections to the coils are felled, the latter is protected against induction voltages (which would damage the coil).

---

[1] Notice: It was not measured if all switches, and especially the switch for coil V_8 and V_9, quench the currents simultaneously

[2] The 0.5A fuse turned out to withstand even 1A for minutes (!?) , but releases immediately at 1.3A. As this is even well below the test currents, this seems to be a proper fuse.

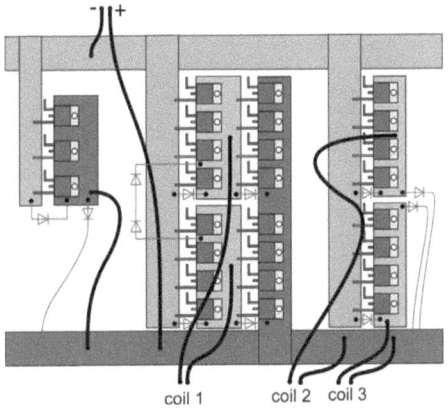

Figure B.2.: Scheme of the MOSFET-benches: short-circuit (left), H-bridge (middle) and two one-way switches (right) are illustrated

B. Electronics

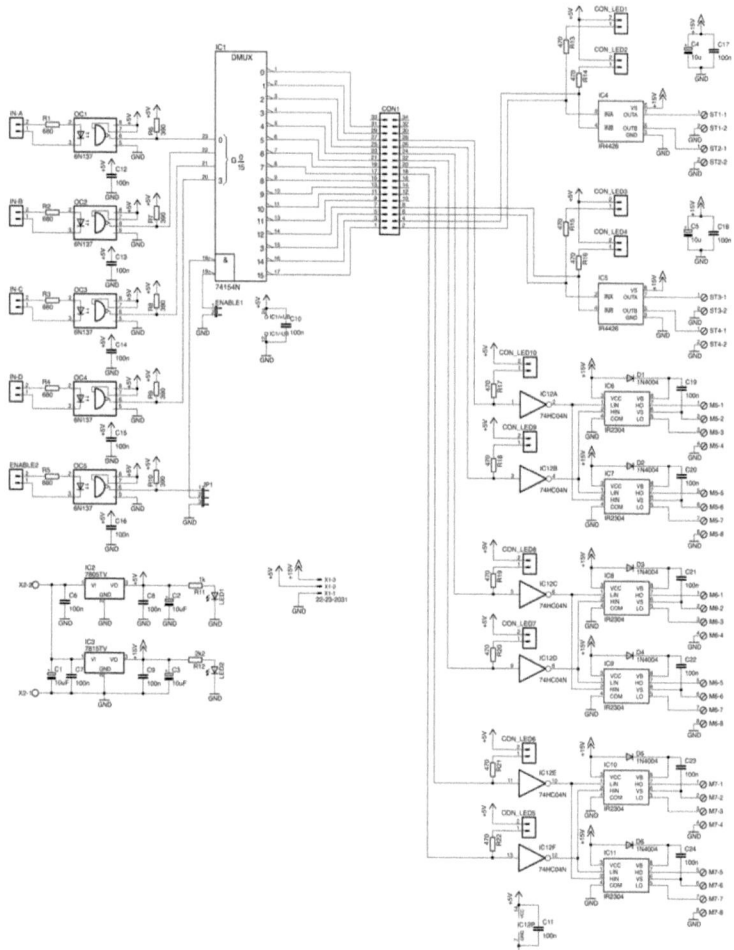

Figure B.3.: Circuit diagram of the de-multiplexer board

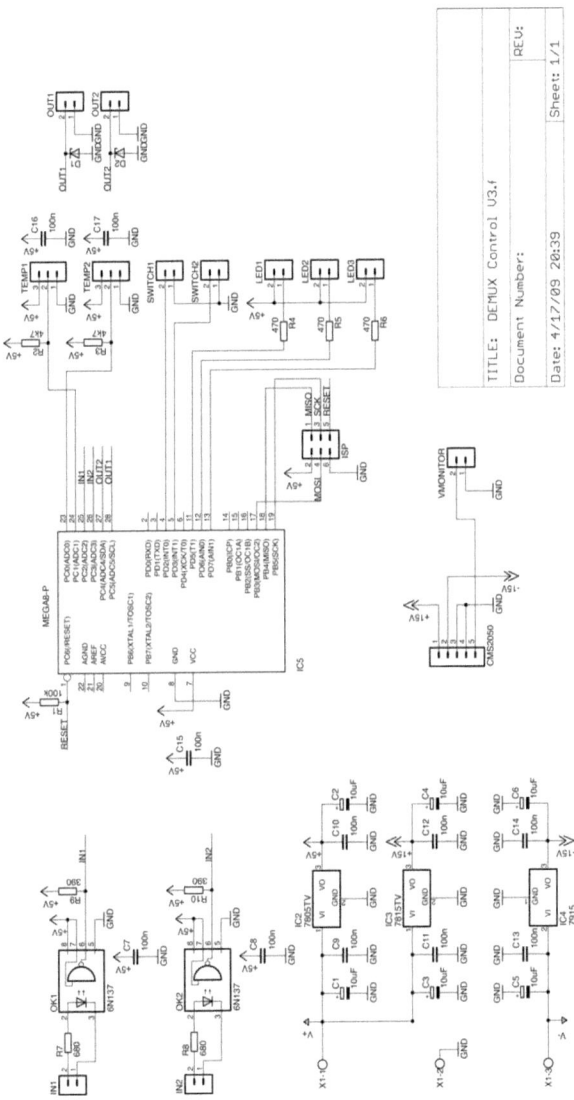

Figure B.4.: Circuit diagram of the demux-control board

B. Electronics

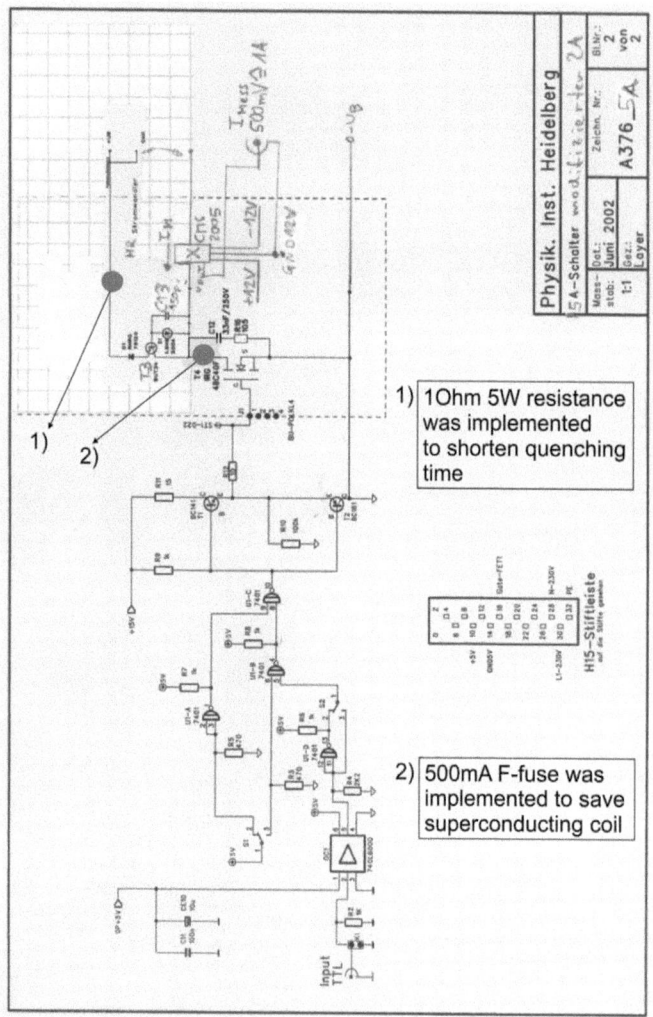

Figure B.5.: Circuit scheme of the superconducting coil switch, with the adaption implemented. 1) a serial resistance to shorten the switch-o time of the coil, and 2) a fuse to protect the coil against to high currents.

C Experimental control

Experimental control for a magnetic transport into a cryostat

To control the experiment, a real-time processing control is implemented based on the PROII-system from Jäger Messtechnik[1]. It provides 64 digital I/O channels and 32 analog output channels which are used to drive the power-supplies. Fig.(C.1) illustrates the setup exhibiting the main parts of the control-setup.

Changes to the transport protocols and to ADwin

The general explanation for the ADwin experimental-control can be found in [178]. In addition, the control has been extended to 64 digital-out channels, while the analog-out channels were extended from $24 \to 32$.

The currents for the transport are calculated from a file which is called *stromfits.m* and are than recalculated as $I(x) \to I(t)$. The resolution of $I(x)$ is $0.1\mu m$ which leads to 2104 entires in the horizontal part and to 2150 entries in the vertical current vectors. Each current curve is read in, converted to $I(t)$, using the defining parameter t, v, a and sliced into 16 parts which are separately fitted with cubic splines defined by

$$I(t) \propto a_0 + ... + a_{n-1}(t - t_{start})^{n-1} \tag{C.1}$$

for $n = 2, 3, 4$. These cubic splines then defines in the corresponding regions the current I(t) as it is then used for processing in the ADwin. A single time-step corresponds to the internal timer-duration $\Delta t = 25\mu s$. This value was changed as $\Delta t = 20\mu s$ turned out to conflict with 32 Analog-Out channels.

---

[1] Jäger Computergesteuerte Messtechnik GmbH D-64653 Lorsch

C. Experimental control

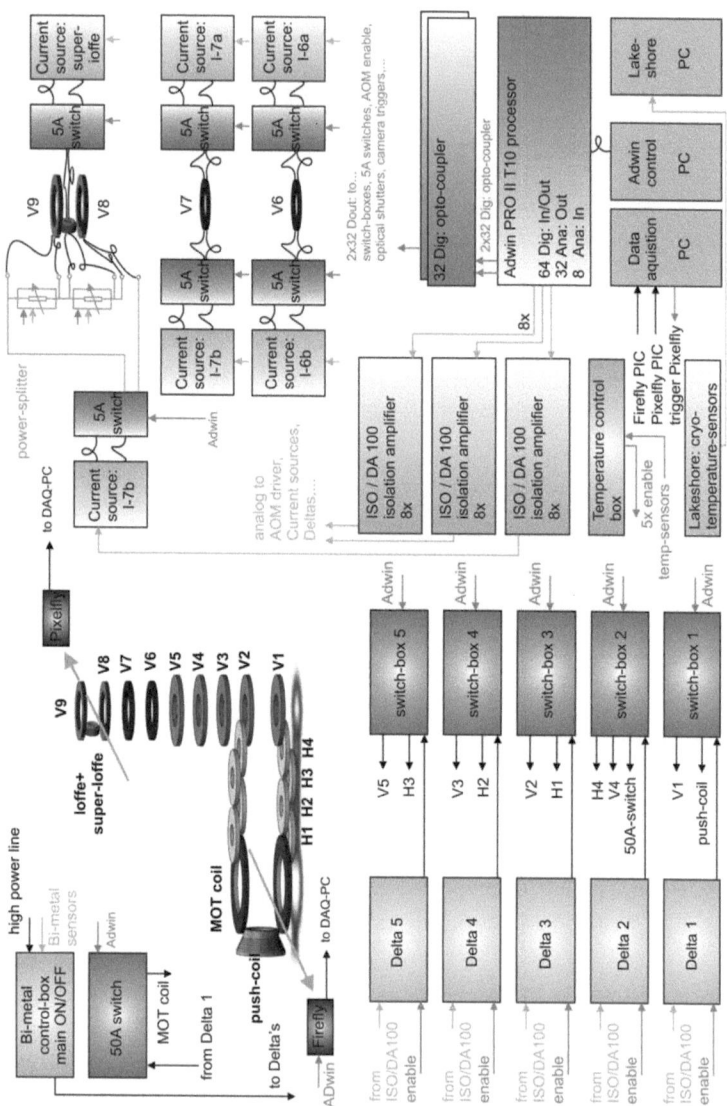

Figure C.1.: Experimental control setup illustrating the power-supplies, switches and switch-boxes and how they are controlled by ADwin

Figure C.2.: Flow chart for the experimental control

D Material Properties

The values are taken from [182, 183] and additionally from other sources for the different materials: Steel 316LN [219, 220], Aluminum $Al99.5\%$ [221], GeVarnish[1] [222], Stycast[2], Apiezon[3], SF57[4] Kapton[5]

Metals	Ther. Cond.	Specific Heat C_V	Elect. Resist. $_{el}$	Temp. T
	$[W/m\ K]$	$[J/g\ K]$	$[\frac{mm^2}{m}]$	[K]
Steel 316LN	14.7	0.48	0.75	295
	7.5	0.19	0.566	55
	0.25	0.0019	0.539	4
Al 99.5%	220	0.902	0.0290	300
	≈ 320	≈0.15	≈0.003	55
	55	0.0025	≈0.0008	4.2
Cu RRR= 50	≈ 350	0.385	0.017	300
	≈ 800	0.1	0.00045	55
	≈ 300	0.00008	$1/50\ R_{300}$	4
Cu RRR= 500 (OFHC)	≈ 350	0.385	0.017	300
	≈ 1000	0.1	0.00045	55
	≈ 3000	0.00008	$1/500\ R_{300}$	4

Table D.1.: Thermal and electrical properties of metals

---

[1] Technical information from: CMR-direct, Cambridge Magnetic Refrigeration, 19-21 Godesdone Road, Cambridge, CB5 8HR; and from from Lake Shore Cryotronics, Inc.
[2] Technical information from: Emerson and Cuming, Billerica, MA 01821 - USA
[3] Technical information from: Lake Shore Cryotronics, Inc.
[4] Technical information from: Advanced Optics SCHOTT AG, 55122 Mainz - D
[5] Technical Information from: DuPont High Performance Films, Circleville, OH 43113 - USA

D. Material Properties

Glue & Grease	Ther. Cond.	Specific Heat	Elect. Resist.	Vap. Press.	Temp.
		C_V	el	p_v	T
	$[W/m\ K]$	$[J/g\ K]$	$[\frac{mm^2}{m}]$	[mbar]	[K]
Stycast	1.3	≈1.3	$5\ 10^{20}$	< 0.125	300
	≈ 0.3	≈0.083	—	—	55
	0.064	≈0.0005	—	—	4
GeVarnish	0.44	1.3	$1.014\ 10^7$	rel. high	300
IMI 7031	0.22	—	—	rel. high	77
	0.062	—	—	—	4.2
Apiezon N	0.194	> 1.2	2×10^{22}	2.7×10^{-9}	300
	0.11	0.657	—	—	100
	0.005	0.00203	—	—	4

Table D.2.: Thermal and electrical properties of Glues & Grease

Chip mountings	Ther. Cond.	Specific Heat	Temp.
		C_V	T
	$[W/m\ K]$	$[J/g\ K]$	[K]
Sapphire Al_2O_3	47	≈ 0.779	295
	450	0.126	100
	10000	≈ 0.015	55
	2900	0.00009	10
	230	—	4
SiO_2 quartz-crystal	9	≈ 0.745	295
	43	0.261	100
	1345	0.0007	10
	185	—	4
MgO-crystal	61	≈ 0.940	295
	507	0.208	100
	1130	< 0.002	10
	82	—	4
Window glass			
SF57	0.62	360	300
	—	—	55
	—	—	4

Table D.3.: Thermal properties for chip mountings

	Ther. Cond.	Specific Heat	Elect. Resist.	Temp.
		C_V	ρ_{el}	T
	$[W/m\ K]$	$[J/g\ K]$	$[\Omega\ cm]$	[K]
Kapton	0.15	0.755	2.3×10^{16}	300
	0.09	≈0.224	—	55
	0.005	0.00079	—	4
Teflon	0.27	0.87	—	295
	0.23	≈0.31	—	77
	0.046	< 0.026	—	4
Solder (60Sn40Pb)	0.173	—	15	295
	53	—	3	77
	16	—	SC	4

Table D.4.: Thermal and electrical properties of used materials

E Rubidium data

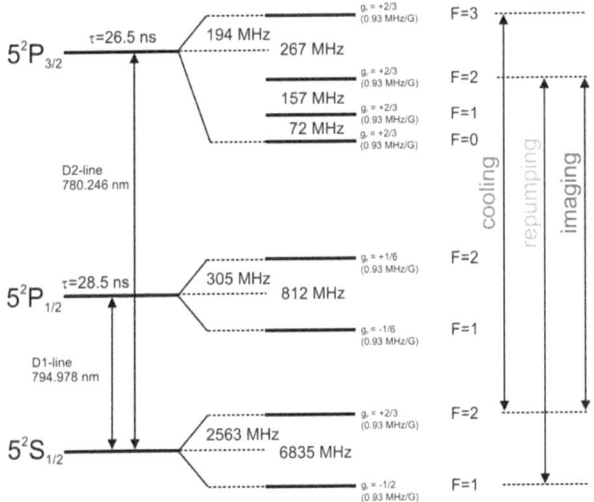

Figure E.1.: Relevant transitions of ^{87}Rb

E. Rubidium data

Atomic Number	Z	37
Total Nucleons	Z+N	87
Relative Natural Abundance	^{87}Rb	27.83(2)%
Nuclear Lifetime	n	4.88 · 10^{10} yr
Atomic Mass	m	86.909 u
Density at 25°C	m	1.53 g/cm^3
Melting Point	T_M	39.31
Specific Heat Capacitance	c_p	0.363 $J/g \cdot K$
Molar Heat Capacitance	C_p	31.060 $J/mol \cdot K$
Vapor Pressure at 25°C	p_v	4.0 · $10^{-7} mbar$
Nuclear Spin	I	3/2
Ionization Limit	E_I	4.1771370(2) eV

Table E.1.: Physical properties of ^{87}Rb, taken from [140].

Wavelength in vacuum		780.246291nm
Frequency	ω_0	2π · 384.227981877THz
Lifetime		26.24ns
D2-saturation Intensity	I_S	1.669(2)mW/cm^2
Natural Line-width	$\Gamma = 1/\tau$	2π · 6.065MHz
Decay Rate in vacuum		38.11(6) · $10^6 s^{-1}$
Recoil Velocity	$v_R = \frac{\hbar k}{m}$	5.8845 mm/s
Recoil Temperature	$T_r = \frac{1}{3} \frac{\hbar^2 k^2}{m k_B}$	361.95nK
Doppler Temperature	$T_D = \frac{\hbar \Gamma}{2 k_B}$	145.5μK

Table E.2.: Optical properties of ^{87}Rb, and important deduced parameter for the D2-line ($5^2 S_{1/2} \to 5^2 P_{3/2}$), taken from [140].

F Physical constants

Speed of Light	c	$2.99792458 \times 10^8 \ m/s$
Permeability of vacuum	μ_0	$4\pi \times 10^{-7} \ N/A^2$
Planck's Constant	h	$6.62606876(52) \times 10^{-34} \ J \ s$
Avogadro constant	N_A	$6.02214179(30) \times 10^{23} \ mol^{-1}$
Boltzmann's Constant	k_B	$1.3806503(24) \times 10^{-23} \ J/K$
Elementary Charge	e	$1.602176462(63) \times 10^{-19} \ C$
Bohr Magneton	μ_B	$9.27400899(37) \times 10^{-24} \ J/T$
Atomic Mass unit	u	$1.66053873(13) \times 10^{-27} \ kg$
Electron Mass	m_e	$5.485799110(12) \times 10^{-4} \ u$

Table F.1.: Physical Constants

Permittivity of vacuum	$\epsilon_0 = \frac{1}{\mu_0 c^2}$	$8.854187817 \times 10^{-12} \ V/(A \ m)$
Ideal Gas constant	$R = N_A \times k_B$	$8.314472(15) \ J/(K \ mol)$
Stefan-Boltzmann Constant	$\sigma = \frac{2\pi^5 k_B^4}{15 h^3 c^2}$	$5.670400(40) \times 10^{-8} \ \frac{W}{m^2 K^4}$

Table F.2.: Deduced Physical Constants

G Construcion Drawings

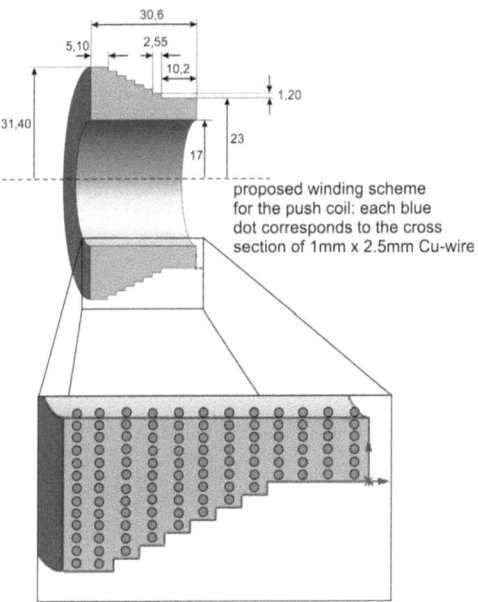

Figure G.1.: Technical Construction Drawing: Push-coil winding scheme

G. *Construcion Drawings*

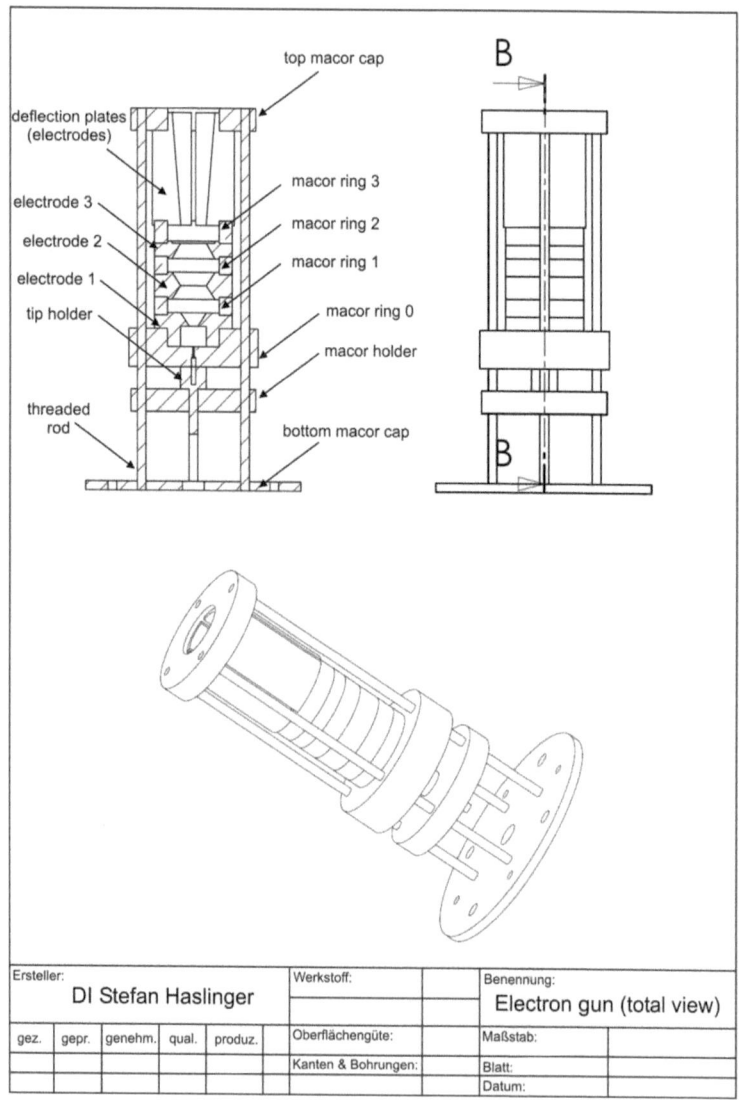

Figure G.2.: Technical Construction Drawing E-gun: Overview of the electron-gun

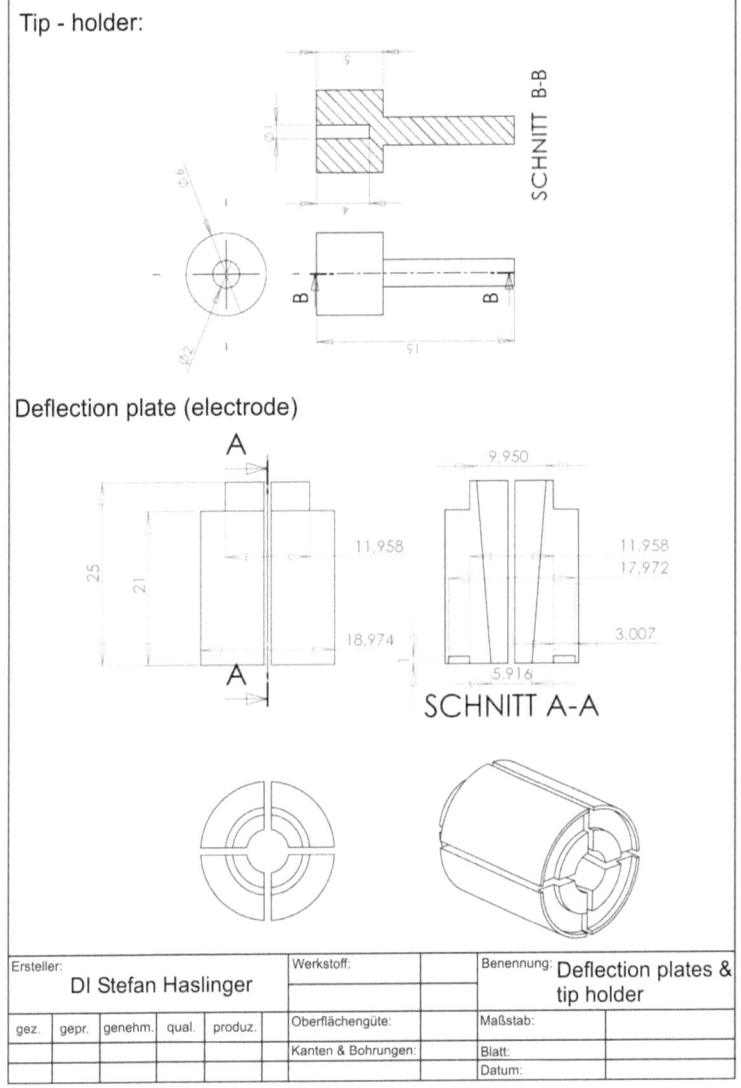

Figure G.3.: Technical Construction Drawing E-gun: Field emission tip-holder and deflection plates

G. Construcion Drawings

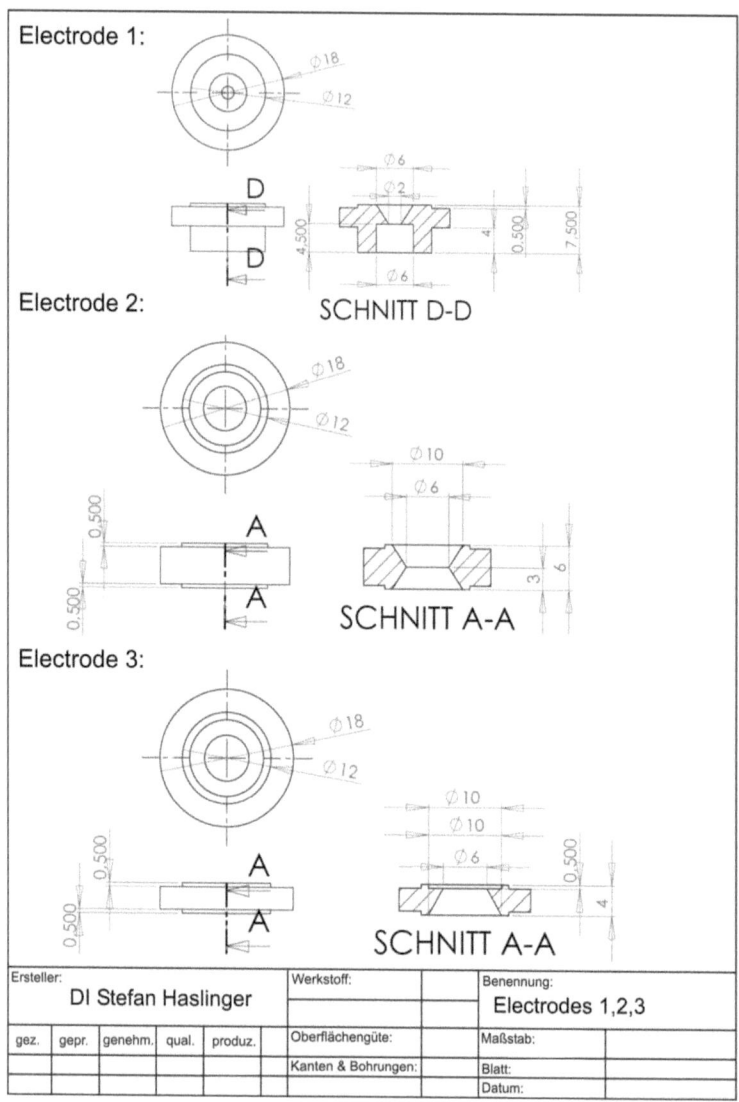

Figure G.4.: Technical Construction Drawing E-gun: Electrodes of the lens-system

240

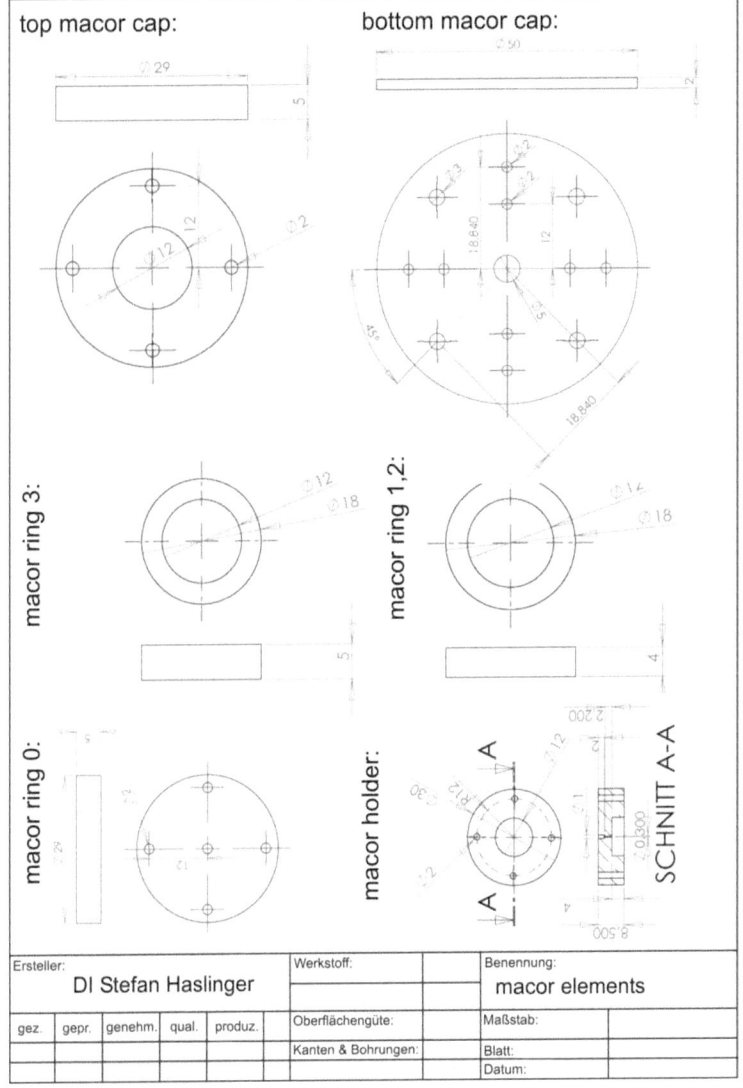

Figure G.5.: Technical Construction Drawing E-gun: Various macor-insulators

G. Construcion Drawings

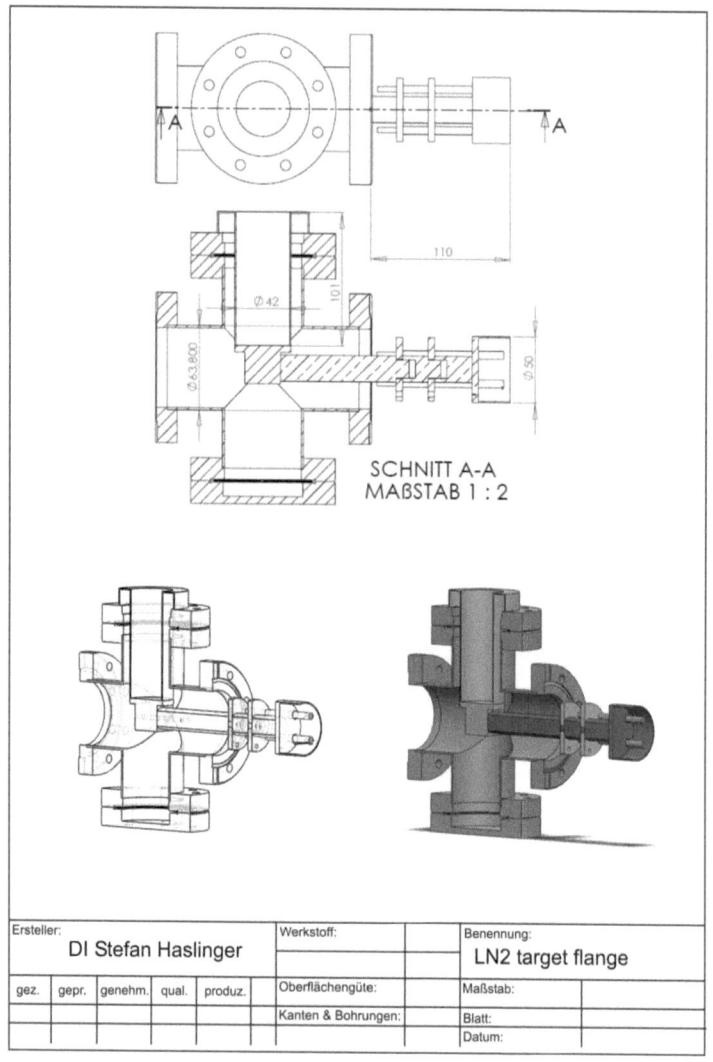

Figure G.6.: Technical Construction Drawing E-gun: Overview of the inner life of the LN_2 target flange

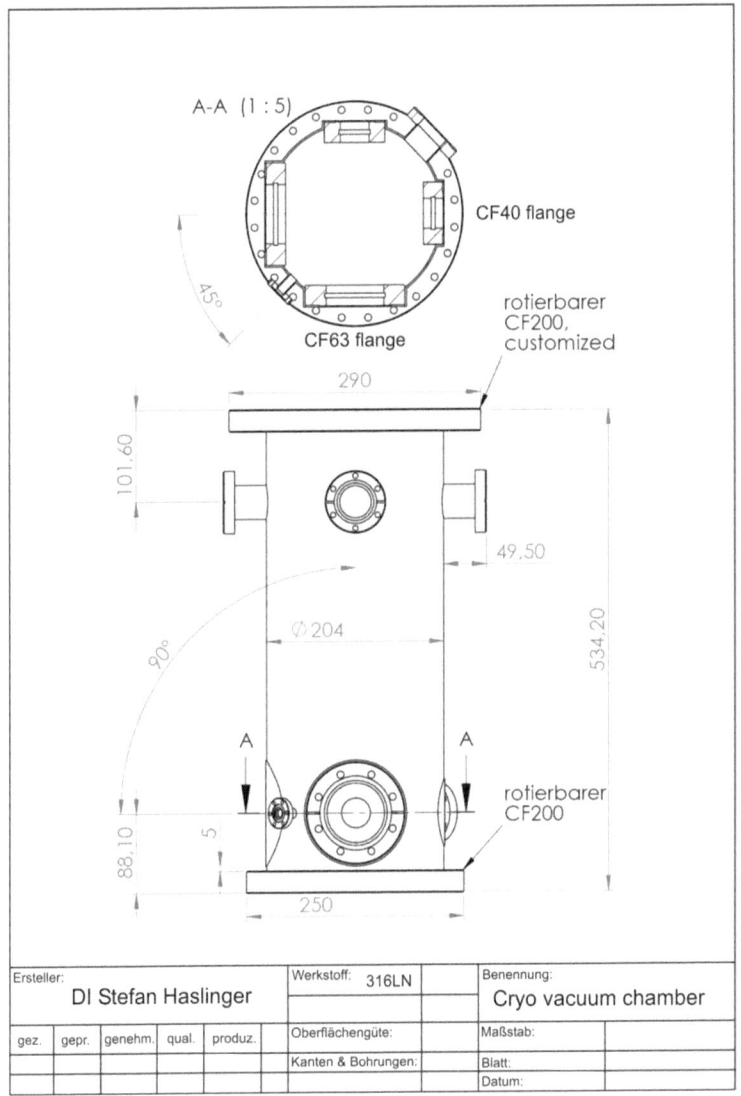

Figure G.7.: Technical Construction Drawing: Cryostat vacuum-chamber

G. Construcion Drawings

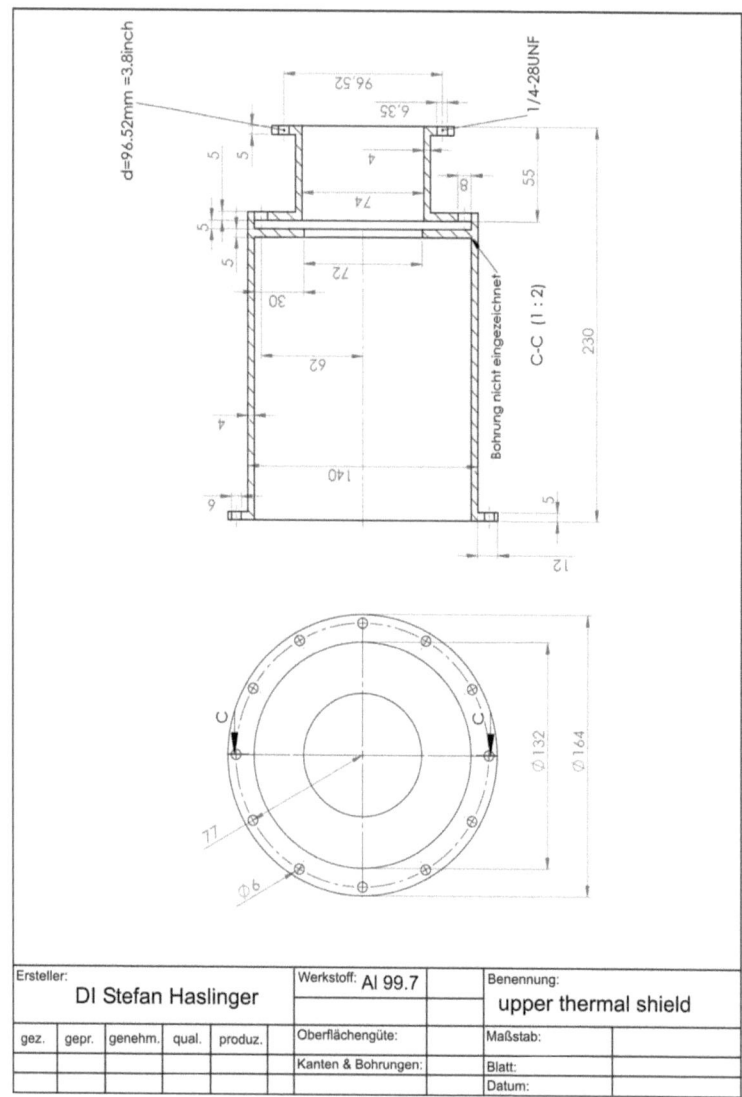

Figure G.8.: Technical Construction Drawing: Upper thermal shield

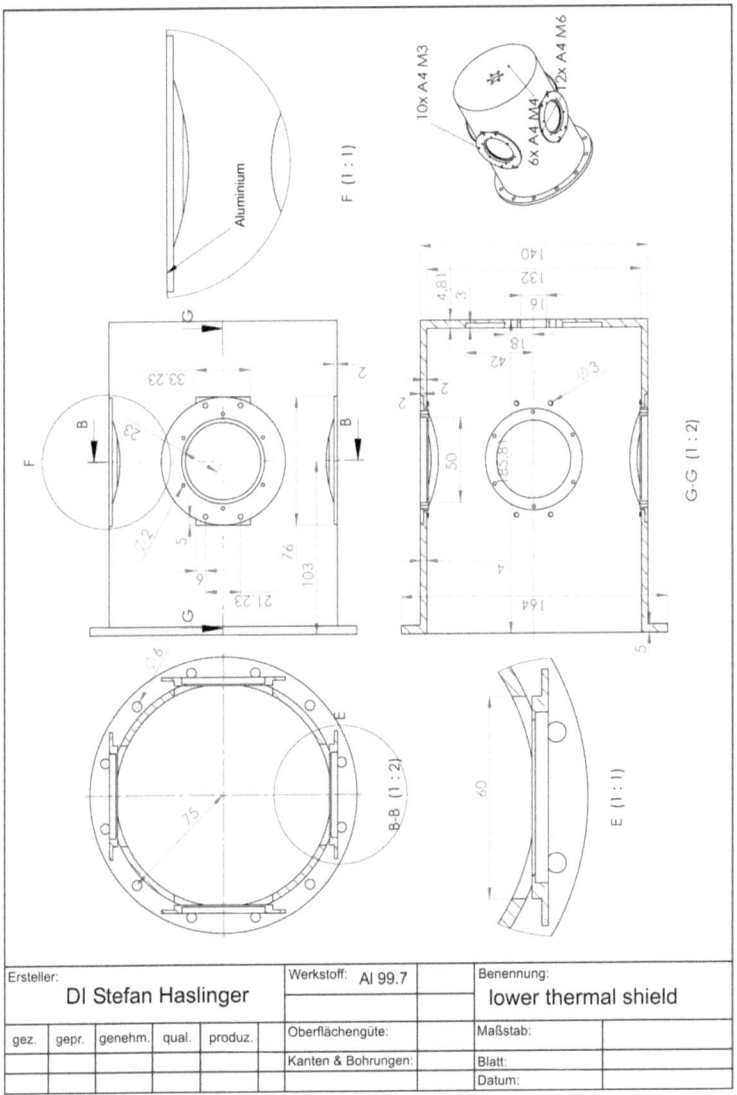

Figure G.9.: Technical Construction Drawing: Lower thermal shield

G. Construcion Drawings

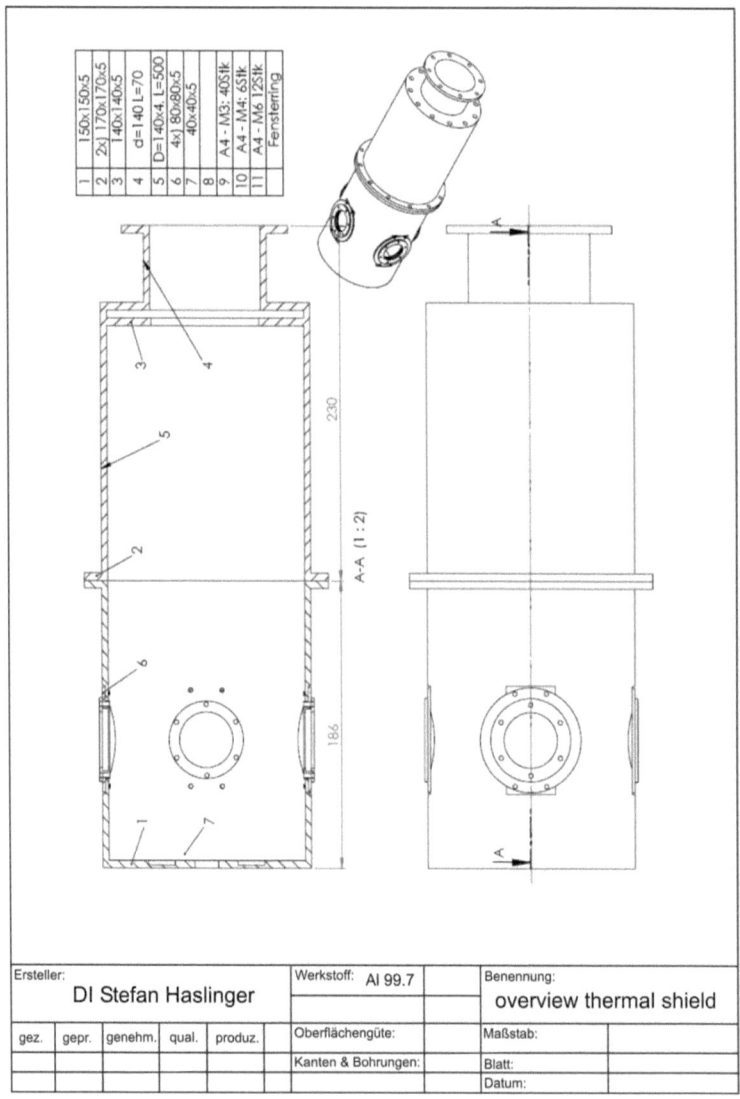

Figure G.10.: Technical Construction Drawing: Overview of the thermal shielding at the 1st-stage

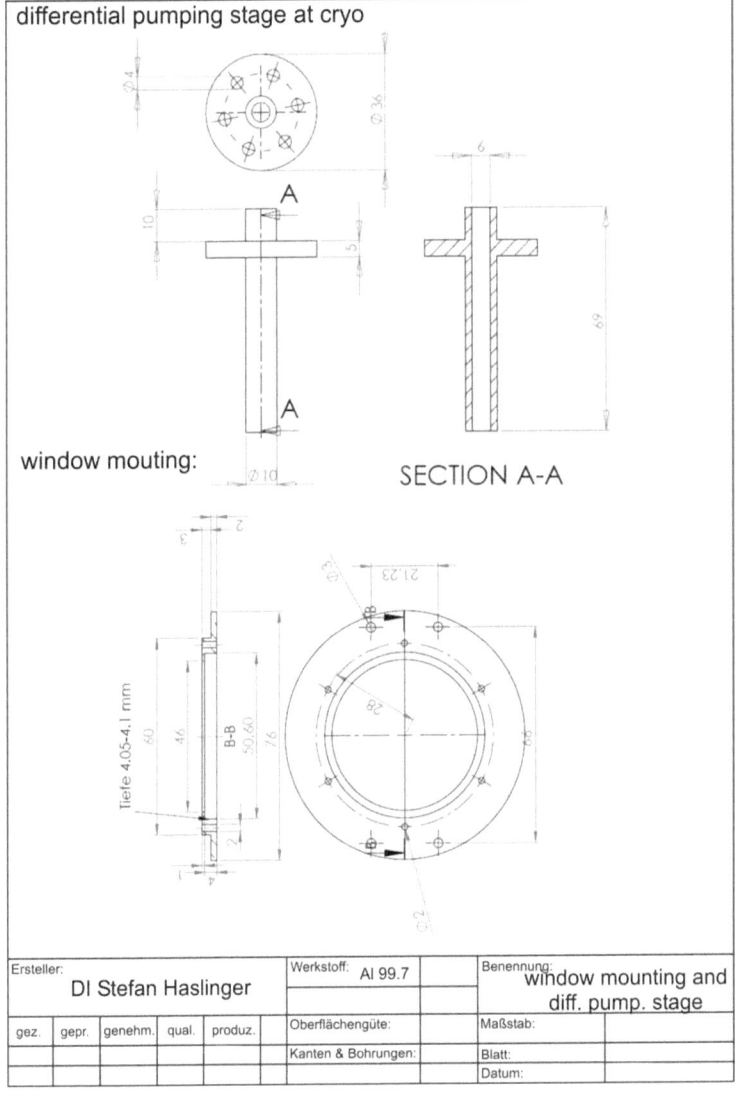

Figure G.11.: Technical Construction Drawing: 1st-stage window mounting (99.5 Al)

G. Construcion Drawings

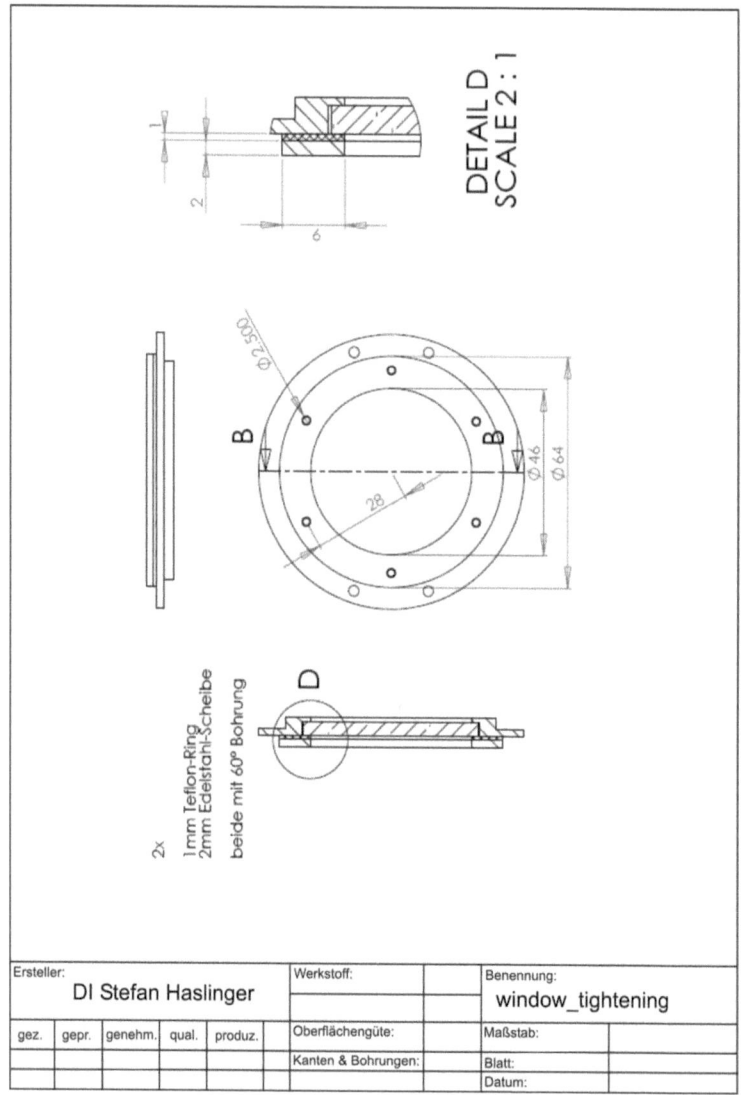

Figure G.12.: Technical Construction Drawing: 1st-stage window tightening steel-ring

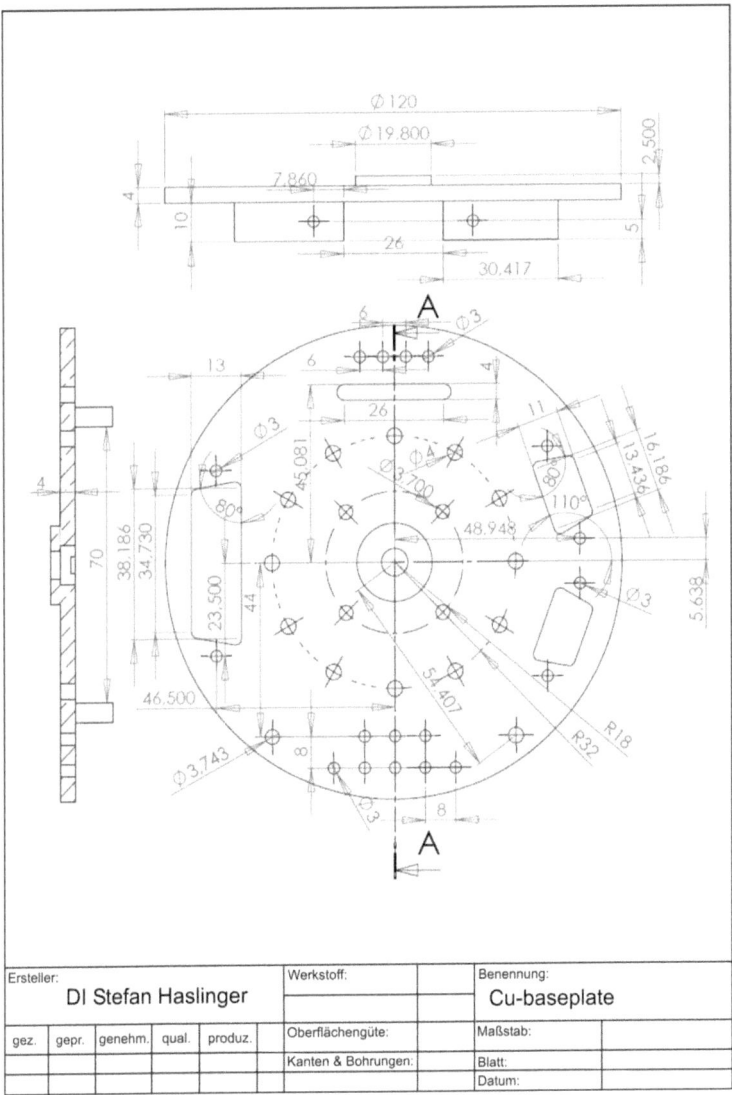

Figure G.13.: Technical Construction Drawing: Cu-base to sustain the 4K experimental setup

G. *Construcion Drawings*

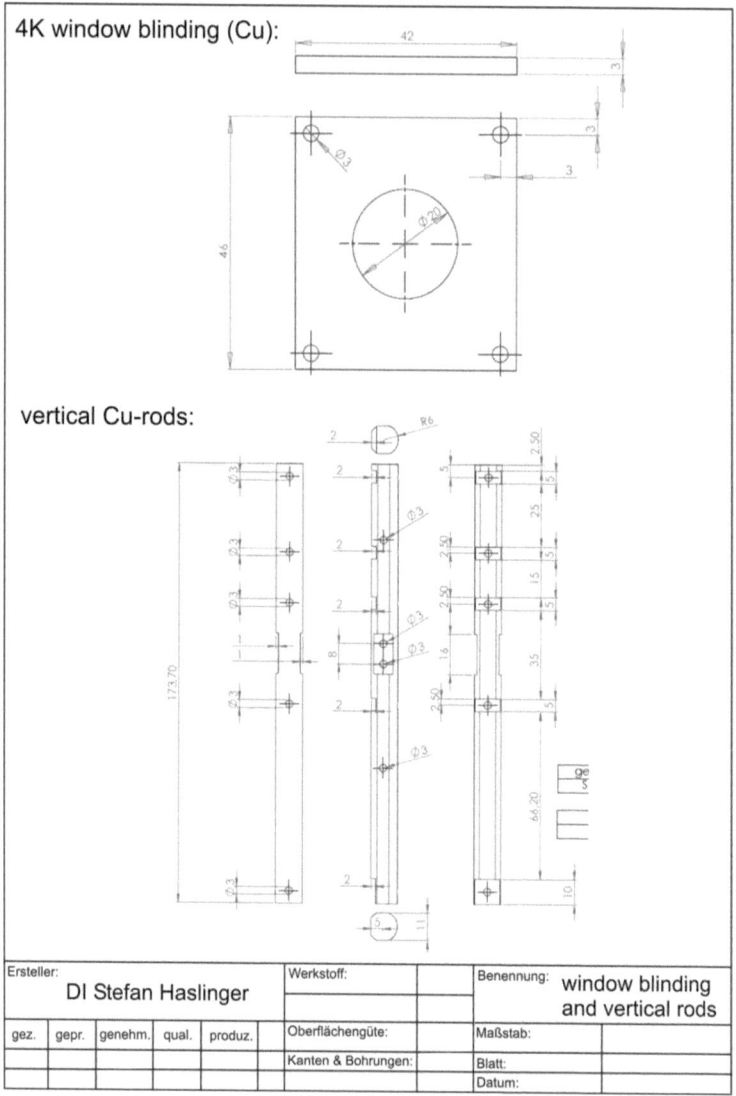

Figure G.14.: Technical Construction Drawing: Vertical Cu-rods for sustaining the coil-cage

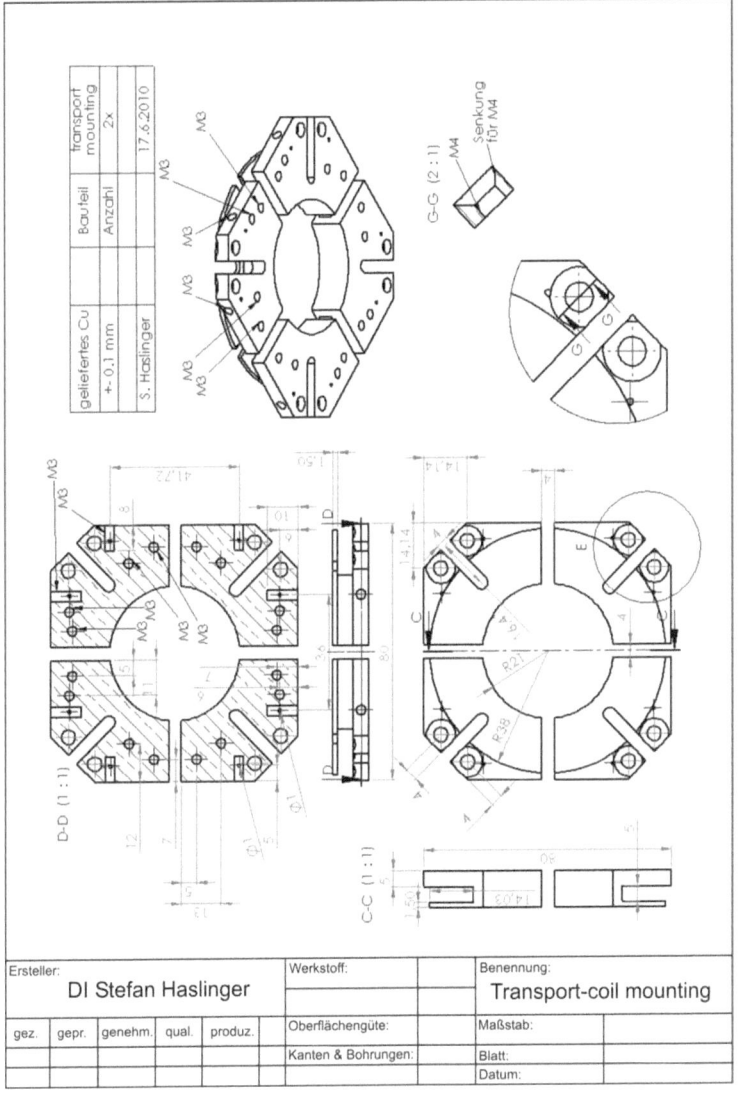

Figure G.15.: Technical Construction Drawing: Transport-coil mounting

G. Construcion Drawings

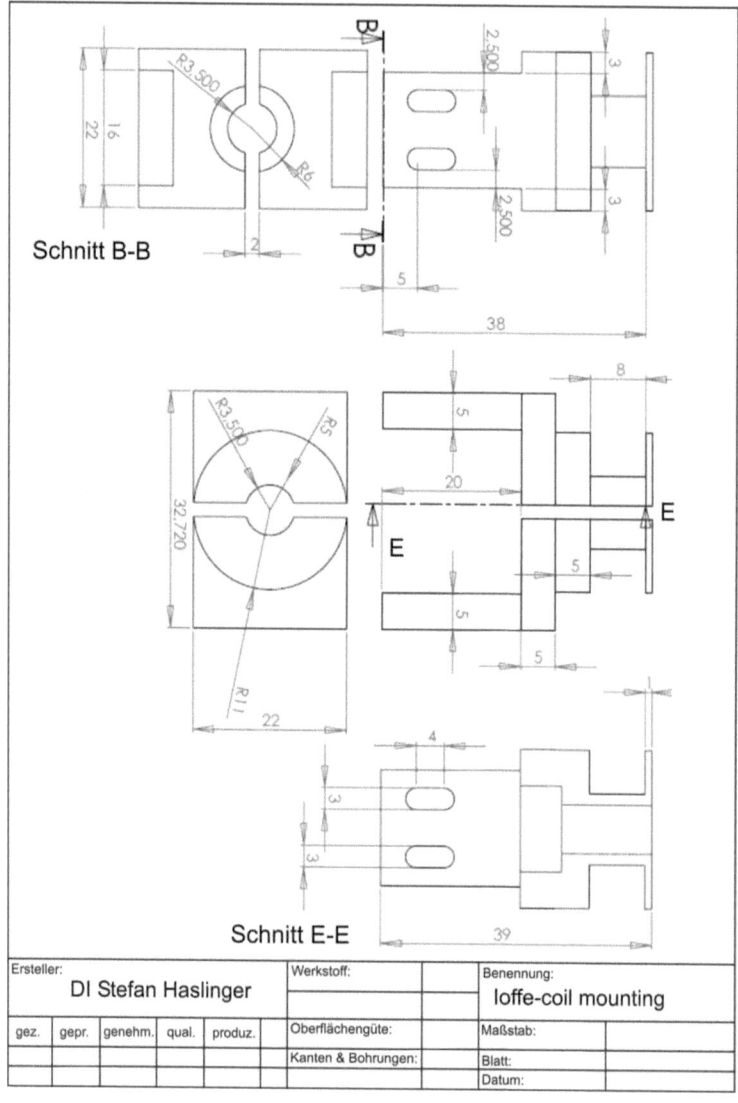

Figure G.16.: Technical Construction Drawing: Ioe-coil mounting

Electron beam driven alkali metal atom source for loading a magneto-optical trap in a cryogenic environment

S. Haslinger · R. Amsüss · C. Koller · C. Hufnagel · N. Lippok · J. Majer · J. Verdu · S. Schneider · J. Schmiedmayer

Received: 6 September 2010 / Revised version: 20 December 2010
© Springer-Verlag 2011

Abstract We present a versatile and compact electron beam driven source for alkali metal atoms, which can be implemented in cryostats. With a heat load of less than 10 mW, the heat dissipation normalized to the atoms loaded into the magneto-optical trap (MOT) is about a factor 1000 smaller than for a typical alkali metal dispenser. The measured linear scaling of the MOT loading rate with electron current observed in the experiments indicates that electron stimulated desorption is the corresponding mechanism to release the atoms.

Preparing ensembles of ultra-cold atoms with a magneto-optical trap (MOT) has become a standard technique in atomic physics and is the first major step on the way to Bose–Einstein condensation (BEC) [1, 2] and ultra-cold quantum gases. Typical sources for alkali metal atoms used in BEC experiments [3, 5–8] are, among others: alkali metal ovens [4] feeding Zeeman slowers [9] or alkali metal dispensers [10] generating vapor of the alkali atoms from where the MOT is loaded. The latter use resistive heating

S. Haslinger (✉) · R. Amsüss · C. Koller · C. Hufnagel ·
N. Lippok · J. Majer · J. Verdu · S. Schneider · J. Schmiedmayer
Vienna Center for Quantum Science and Technology,
Atominstitut, TU-Wien, 1020 Vienna, Austria
e-mail: haslinger@ati.ac.at
url: http://www.ati.ac.at

N. Lippok
Planet and Star Formation Department, Max-Planck-Institut für Astronomy, 69117 Heidelberg, Germany

J. Verdu
Department of Physics and Astronomy, University of Sussex, Brighton, UK

Published online: 02 March 2011

to chemically reduce compounds of alkalies to produce the atomic vapor.

Recently, the increasing interest for studying the interaction between ultra-cold quantum gases and solid state quantum devices [11–15] brought up a need for alkali metal sources compatible with the thermal load in cryogenic environments where superconducting quantum devices operate. As the ideal platform for such experiments is an atom chip [16–19], state-of-the-art experiments with ultra-cold atoms on superconducting atom chips use sophisticated transport schemes [20–25]. Implementations are realized by transferring pre-cooled trapped atoms into the cryostat using a moveable magnetic trap [21], optical tweezers [26] or pushing a slow beam of atoms into a cryogenic system with an integrated MOT [25].

In the following, we present a versatile and compact electron beam driven cold atom source which is compatible with a cryogenic environment. Depending on the preparation of the field-emission tip, it is possible to set free a significant amount of trappable atoms at a few hundreds μW of heat load. We observe MOT loading rates which, when scaled by total heat load to the system, are a factor of 1000 higher than for a typical MOT loaded by an alkali metal dispenser. In a typical experiment, we load 3×10^6 atoms within 1.5 seconds at a total heat load of 8.4 mW. Compared with other atom sources conceived for cryogenic systems, such as light-induced atomic desorption (LIAD) [27, 28] or laser ablation [29], this atom source is fully tunable as it relies on an electron beam, rather than on laser light as input power source.

The setup of an electron beam driven atom source is shown in Fig. 1. An electron beam is created by a field-emission source. The electron beam is then directed onto a cryogenic Rb target at 77 K. Using PtIr or tungsten field-emission tips, we achieve a high flux of laser-trappable ru-

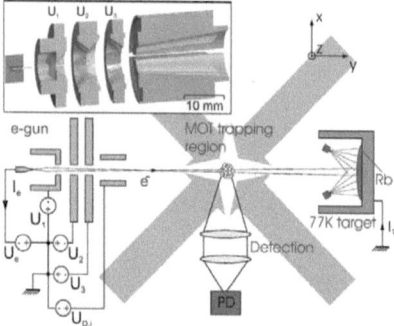

Fig. 1 Schematic of the cryogenic e-beam driven alkali atom source for loading a *MOT*: The electron beam emerges from a cold field-emitter, crossing the trapping region with the crossed laser beams and hitting a liquid nitrogen cooled rubidium target. Two alkali metal dispensers are used to prepare an alkali metal layer on the surface. Released ^{87}Rb atoms are loaded into the *MOT* and the fluorescence of the trapped atoms is measured with a photodiode (*PD*). The *inset* shows a 3D image of the conical deflection plates and the high efficient cold field-emission source. The kinetic energy of the electrons is given by the emission voltage U_e

bidium atoms at maximum target heat loads of < 10 mW. The Rb atoms desorbed from the target by the electron impact are trapped in a close by magneto-optical trap, situated in front of the target at a distance of 7 cm.

The electron beam is prepared by a field-emission tip which avoids thermal radiation from the electron source. We used both PtIr or tungsten tips. With either of the two tips, the source is capable of producing a beam current of more than 10 µA at kinetic energies up to 6 keV. To preserve the tip, the field-emitter is operated at currents below 10 µA.

With a system of electrostatic lenses, the electron beam is focused onto the target to a spot size of ≤ 600 µm diameter. The design of the lens system is based on [36, 37], where particular attention was paid to optimize efficient transmission via minimizing the loss of the emitted current I_e propagating through the lens system. The transfer efficiency I_T/I_e of the target current I_T and the emitted current is up to 0.4. The other electrons hit the electrodes of the lens system and do not reach the target.[1] We characterized the electron beam in previous experiments, where the target was replaced with a phosphor screen. Measuring the beam profile, beam position and beam current, we used the deflection plates to minimize the effect of the magnetic quadrupole field of the MOT on the electron beam. In addition, the four conical deflection

plates allow the beam to be adjusted in x- and z-directions up to 130 mrad (see Fig. 1), applying a voltage of $U_{D,i=1..4}$ at each of the plates. The target current I_T and the emission current I_e are continuously monitored.

Leaving the field-emission source, the electrons cross the MOT region and hit the liquid nitrogen cooled target. For the used kinetic energies of the electrons, the short loading times and the low atom density, the probability of ionizing the Rb [38, 39] in the MOT is much less than 1% and can therefore be neglected.

Initially our Rb target is prepared the following way: Rb is deposited on the oxidized Cu surface from two very close by (see Fig. 1) alkali metal dispensers. The dispensers [10] are switched on for 60 seconds with 2 × 5A, and coat the LN_2-cooled target surface with a thick Rb film as most of the Rb sticks on the Cu surface (partial pressure of Rb at 77 K is about 10^{-15} mbar [30]). The target lasts for more than 1000 experimental cycles.

To characterize the vacuum during target preparation and the experimental cycle, we observe the overall pressure in the chamber with an UHV gauge and the partial pressure of ^{87}Rb and other rest gas species using a mass spectrometer. During a typical experimental cycle, the total pressure is < 2×10^{-10} mbar.

When the electrons hit the target, neutral rubidium atoms are desorbed and the low velocity tail of the released ^{87}Rb atoms can be trapped and laser-cooled in the MOT. We operate the MOT using a diode laser, $\delta = -18$ MHz detuned from the $5^2S_{1/2}$ (F = 2) → $5^2P_{3/2}$ (F' = 3) transition at 780 nm and a typical quadrupole gradient of 20 G/cm. A second diode laser tuned to the $5^2S_{1/2}$ (F = 1) → $5^2P_{3/2}$ (F' = 2) transition pumps atoms back from the F = 1 to the F = 2 hyperfine state. In each of the three retro-reflected trapping beams we employ ≈20 mW of laser power. At the end of each experimental phase the MOT laser frequency is ramped through resonance ($\delta = 0$) to 5 MHz blue detuning within 2 ms. The number of collected atoms can be calculated via fluorescence from the current peak value of the photodiode at resonance $\delta = 0$, where the signal of the calibrated photodiode is proportional to the number of trapped atoms [31]. Ramping the detuning into the blue, expels all atoms from the trap and resets the number of trapped atoms to zero.

Figure 2 shows the experimental cycle which consists of phases I–IV. We characterize our Rb source by means of a measurement cycle where the MOT is loaded in each phase, lasting 1500 ms.

In phase I, the MOT is loaded without the electron beam switched on, in order to get a reference measurement of the background.

In phase II the electron beam is switched on and the desorbed atoms are trapped in the MOT. At the end of phase II the electron beam is switched off by setting the emission voltage to zero.

---

[1] As simulation software for the optimization of the electron gun and the lens system, the COMSOL Multiphysics Modeling and Simulation package was used.

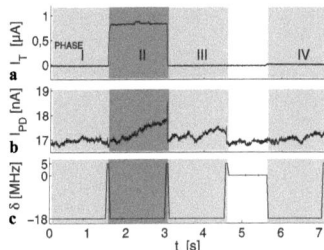

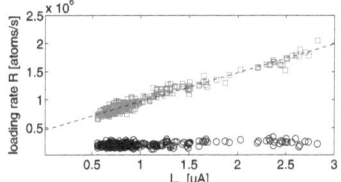

Fig. 2 (a) Electron beam current at target. (b) Fluorescence signal I_{PD} from the photodiode, from which the number of trapped atoms is determined. The signal shows resonance peaks after 1.5, 3, 4.6 and 7.1 seconds, when the cooling laser is ramped over resonance. (c) At the end of every MOT phase, lasting 1.5 seconds, the 780 nm cooling laser at the $5^2S_{1/2}$ ($F = 2$) → $5^2P_{3/2}$ ($F' = 3$) transition is ramped from $\delta = -18$ to $\delta = +5$ MHz within 10 ms. During the pause between phases III and IV, the laser stays at resonance $\delta = 0$

Fig. 3 A typical measurement of the loading rate (□) in dependence on the target current I_T. The atoms are desorbed during phase II, and loaded with a yield Y_D proportional to the slope of the dashed line. A reference measurement (○) from phase I shows the rate of loading the MOT from the background

During phase III, we measure how many atoms are loaded from the remaining background gas while the e-gun is off, where N_{III} is in the order of $< 0.4 \times N_{II}$.

In phase IV, we switch the electron emitter on, but block the electron beam such that it cannot reach the target. This is achieved by applying a large negative voltage U_3 at the blocking electrode, which is ramped up between phase III and phase IV. Technical limitation for the ramping speed of the blocking voltage leads to a pause between phase III and phase IV. In phase IV the beam is not hitting the target, even though a small leakage current remains. This phase therefore allows us to conclude that the trapped ^{87}Rb in the MOT originates from the target rather than being released from other surfaces accidentally hit by electrons.

After phase IV, the cycle is paused for 13 seconds before restarting with phase I. The total experimental cycle lasts 20 seconds.

During the operation of the electron gun the first electrode is biased with $U_1 = -500$ V and the emission voltage U_e is varied from -2.5 up to -3.0 kV to adjust the electron beam current which is related to the emission voltage U_e by the Nordheim–Fowler equation [32]. The focusing voltage U_2 is set to -440 V.

Operated with a PtIr emission tip, we observe that an increase of the electron current on the target I_T results in a linear increase in the number of trapped atoms N_{II} during phase II and hence in the loading rate R (see Fig. 3). In addition, we measure no dependence of the trapped atoms on the electron beam spot size. In a typical experiment, we trap about 3×10^6 atoms at a target current of 2.8 μA after a loading time τ_L of 1.5 seconds and 4.5×10^6 atoms after $\tau_L = 2.5$ seconds. In a different measurement were we load the MOT for up to 20 seconds, we find a characteristic time constant of about 10 seconds, fitting an exponential law to the loading curve of the MOT. This loading time constant is much longer than the loading time $\tau_L = 1.5$ seconds during phases I–IV. This results in a near linear behavior of the loading curve, and the loading rate can be estimated from the number of trapped atoms divided by τ_L. We determine a loading yield of $Y_D = 5.1 \times 10^5$ atoms/s/μA, which gives the number of atoms in the MOT per μA electron target current I_T and which is proportional to the slope of the curve in Fig. 3.

For a tungsten emitter operated at 650 V and using a divergent electron beam with cross section of several square centimeters on the target, we found a loading rate of $R = 7.3 \times 10^5$ atoms/s at a total target power of only 200 μW. A conventional MOT operated in the same chamber under identical laser and vacuum conditions (background pressure of 1.2×10^{-10} mbar), loaded from an Rb-dispenser, yields a loading rate of $\approx 1.9 \times 10^7$ atoms/s using a resistive power of ≈ 18 W.[2] Comparing both, we see that the electron beam Rb source achieves a good yield of trapped atoms even at very low powers. The total heat load from the e-beam driven Rb source, scaled by the MOT loading rate, is more than a factor 1000 lower than for loading from an Rb-dispenser.

To explain the experimental observations, we consider two potential mechanisms to desorb neutral alkali metal atoms from an oxidized metal surface as a result of electron impact: (1) electron stimulated desorption (ESD) and (2) thermal desorption due to electron beam heating. While the first effect relies on the neutralization of adsorbed alkali metal ions by a charge transfer process from the oxidized surface, in the latter case a part of the electrons kinetic energy is converted into heat and leads to thermal evaporation of surface atoms.

---

[2]In addition, we point out that Rb alkali metal dispensers demand a minimum power in the order of at least 8 W to set free a feasible amount of Rb.

According to Ageev et al. who studied ESD of Li, Na, K and Cs from alkali metal layers on oxidized tungsten [33], an incident electron creates a core-hole in the oxygen 2s level which stimulates an intra-atomic Auger-decay and allows for a subsequent neutralization of a positive alkali metal ion. If the positive oxygen ion can capture electrons from the substrate to achieve a negative charge state again, the alkali metal atom will be repelled and desorb as a neutral atom. This desorption process will increase the partial pressure of an adsorbed species j, and can be described by

$$\Delta Q'_j = \left(p_j^1 - p_j^0\right) S_j = \eta_j \frac{I_e}{e} k_B T \qquad (1)$$

as written in [40]. Here, $\Delta Q'_j$ is the differential rate for desorbing particles of species j, p_j^0 and p_j^1 are the steady-state partial pressures before and after electron impact, S_j is the effective pumping speed of species j, η_j the molecular desorption yield, I_e the total electron current, k_B the Boltzmann constant and T the temperature of the target. Equation (1) assumes that the electron current density is low compared to the adatom density. This leads to the observed linear relationship between the desorption rate and the electron current (Fig. 3).

We also established a simple mathematical model for the thermal desorption of a thin layer of Rb due to electron impact. Following a model from Lin [34], the local temperature rise due to the electron beam on the Rb layer can be estimated. With the temperature known, the Langmuir–Knudsen law [35] describes the presented desorbing process and mass flow. Assuming a thermal velocity distribution and a maximum capture velocity of the MOT, we obtain estimates for the number of trapped Rb atoms in the MOT. Due to the rapidly increasing vapor pressure of Rb with increasing target temperature, one expects a highly non-linear, exponentially shaped relationship between the number of trapped atoms and the electron current on the target in contradiction to our observation (Fig. 3).

We deduce from these simple models that the linear relationship between the number of desorbed atoms and the target current is rather described by ESD than a thermal desorption process. In order to substantiate this, further studies e.g. at lower electron energies would be necessary, which was not possible with our electron source. In addition, we would also like to point out that we observe an increase in the partial pressures of rest gas species (H_2, CO_2, and N_2) in our chamber, during operation of the electron source. This can be understood by ESD of non-metals, induced by stray electrons hitting a surface different from our target. Due to the increased background pressure we were constrained to short MOT loading times.

In conclusion, we present an electron beam driven source for Rb atoms desorbed from a 77 K target, which is able to load a magneto-optical trap. The electron beam is emitted either from a PtIr-tip or an etched W-tip where the PtIr-tip is easier to produce and the W-tip demands less emission voltage to be operated. The linear dependence of the MOT loading rate with the electron current impinging on the target depends weakly on the focused spot size of the beam. In addition, the low power density of the divergent beam suggests that electron stimulated desorption and not thermal evaporation is the mechanism to release the atoms from the surface. With its low power needed to operate, the atom source does not present a significant heat load for a cryogenic cold atom experiment. Our electron beam driven Rb source requires a factor 1000 less input power to load a MOT when compared to standard Rb-dispensers under the same conditions. This demonstrates that electron beam driven atom sources can provide several 10^6 trappable atoms in cryogenic environments with low cooling powers.

Acknowledgements We thank T. Juffmann, Universität Wien for supporting us with tungsten emitters, and J. Summhammer and M. Fugger for technical support. This work was supported by the European Union project MIDAS and the Austrian Science Fund FWF. SH acknowledges support from the DOC program of the Austrian Academy of Science (ÖAW), CK from the FUNMAT research alliance, RA and NL from the COQUS doctoral program. JM acknowledges support from the Marie Curie Action HQS.
S. Haslinger and R. Amsüss both contributed equally to this work.

References

1. E.A. Cornell, C.E. Wieman, Rev. Mod. Phys. **74**, 875 (2002)
2. W. Ketterle, Rev. Mod. Phys. **74**, 1131 (2002)
3. M.H. Anderson, J.R. Ensher, M.R. Matthews, C.E. Wieman, E.A. Cornell, Science **269**, 198 (1995)
4. K.B. Hadzibabic, M.-O. Stan, M.R. Dieckmann, N.J. Gupta, D.S. Zwierlein, D.M. Görlitz, W. Ketterle, Phys. Rev. Lett. **88**, 160401 (2002)
5. K.B. Davis, M.-O. Mewes, M.R. Andrews, N.J. van Druten, D.S. Durfee, D.M. Kurn, W. Ketterle, Phys. Rev. Lett. **75**, 3969 (1995)
6. T. Weber, J. Herbig, M. Mark, H.C. Nägerl, R. Grimm, Science **299**, 232 (2003)
7. G. Modugno, G. Ferrari, G. Roati, R.J. Brecha, A. Simoni, M. Inguscio, Science **294**, 1320 (2001)
8. C.C. Bradley, C.A. Sackett, R.G. Hulet, Phys. Rev. Lett. **78**, 985 (1997)
9. T.E. Barrett, S.W. Dapore-Schwartz, M.D. Ray, G.P. Lafyatis, Phys. Rev. Lett. **67**, 3483 (1991)
10. SAES Getters S.p.A, Alkali metal dispenser datasheet, 20151 Milano, Italy (2003)
11. R.J. Schoelkopf, S.M. Girvin, Nature **451**, 664 (2008)
12. J. Verdú, H. Zoubi, C. Koller, J. Majer, H. Ritsch, J. Schmiedmayer, Phys. Rev. Lett. **103**, 043603 (2009)
13. A.S. Sørensen, C.H. van der Wal, L.I. Childress, M.D. Lukin, Phys. Rev. Lett. **92**, 063601 (2004)
14. D. Petrosyan, G. Bensky, G. Kurizki, I. Mazets, J. Majer, J. Schmiedhauer, Phys. Rev. A **79**, 040304 (2009)
15. D. Petrosyan, M. Fleischhauer, Phys. Rev. Lett. **100**, 170501 (2008)
16. R. Folman, P. Krüger, D. Cassettari, B. Hessmo, T. Maier, J. Schmiedmayer, Phys. Rev. Lett. **84**, 4749 (2000)
17. R. Folman, P. Krüger, J. Schmiedmayer, J. Denschlag, C. Henkel, Adv. At. Mol. Opt. Phys. **48**, 263 (2002)

18. J. Reichel, Appl. Phys. B **74**, 469 (2002)
19. J. Fortágh, C. Zimmermann, Rev. Mod. Phys. **79**, 235 (2007)
20. R. Nirrengarten, A. Qarry, C. Roux, A. Emmert, G. Nogues, M. Brune, J.M. Raimond, S. Haroche, Phys. Rev. Lett. **97**, 200405 (2006)
21. T. Mukai, C. Hufnagel, A. Kasper, T. Meno, A. Tsukada, K. Semba, F. Shimizu, Phys. Rev. Lett. **98**, 260407 (2007)
22. B. Kasch, H. Hattermann, D. Cano, T. Judd, S. Scheel, C. Zimmermann, R. Kleiner, D. Kölle, J. Fortágh, New J. Phys. **12**, 065024 (2010)
23. C. Hufnagel, T. Mukai, F. Shimizu, Phys. Rev. A **79**, 053641 (2009)
24. A. Emmert, A. Lupascu, G. Nogues, M. Brune, J.-M. Raimond, S. Haroche, Eur. Phys. J. D **51**, 173 (2009)
25. C. Roux, A. Emmert, A. Lupascu, T. Nirrengarten, G. Nogues, M. Brune, J.-M. Raimond, S. Haroche, Europhys. Lett. **81**, 56004 (2008)
26. D. Cano, B. Kasch, H. Hattermann, R. Kleiner, C. Zimmermann, D. Koelle, J. Fortágh, Phys. Rev. Lett. **101**, 183006 (2008)
27. M. Meucci, E. Mariotti, P. Bicchi, C. Marinelli, L. Moi, Europhys. Lett. **25**, 639 (1994)
28. S.N. Atutov, R. Calabrese, V. Guidi, B. Mai, A.G. Rudavets, E. Scansani, L. Tomassetti, V. Biancalana, A. Burchianti, C. Marinelli, E. Mariotti, L. Moi, S. Veronesi, Phys. Rev. A **67**, 053401 (2003)
29. S.E. Maxwell, N. Brahms, R. deCarvalho, D.R. Glenn, J.S. Helton, S.V. Nguyen, D. Patterson, J.M. Doyle, J. Petricka, D. DeMille, Phys. Rev. Lett. **95**, 173201 (2005)
30. D.A. Steck, Los Alamos Nat. Lab., technical report LA-UR-03-8638 (2008), http://steck.us/alkalidata/rubidium87numbers.pdf
31. E.L. Raab, M. Prentiss, A. Cable, S. Chu, D.E. Pritchard, Phys. Rev. Lett. **59**, 2631 (1987)
32. R.H. Fowler, L.W. Nordheim, Proc. R. Soc. Lond. Ser. A, Math. Phys. Sci. **119**, 173 (1928)
33. V.N. Ageev, Y.A. Kuznetsov, B.V. Yakshinskii, T.E. Madey, Nucl. Instrum. Methods Phys. Res., Sect. B, Beam Interact. Mater. Atoms **101**, 69 (1995)
34. T. Lin, IBM Syst. J. **11**, 527 (1967)
35. I. Langmuir, Phys. Rev. **2**, 329 (1913)
36. K. Kuroda, T. Suzuki, J. Appl. Phys. **45**, 1436 (1974)
37. K. Kuroda, H. Ebisui, T. Suzuki, J. Appl. Phys. **45**, 2336 (1974)
38. T. Gericke, P. Würtz, D. Reitz, T. Langen, H. Ott, Nat. Phys. **4**, 949 (2008)
39. R.S. Schappe, T. Walker, L.W. Anderson, C.C. Lin, Phys. Rev. Lett. **76**, 4328 (1996)
40. H. Tratnik, Electron stimulated desorption of condensed gases on cryogenic surfaces, Ph.D. thesis Vienna University of Technology, 2005

Die VDM Verlagsservicegesellschaft sucht für wissenschaftliche Verlage abgeschlossene und herausragende

Dissertationen, Habilitationen, Diplomarbeiten, Master Theses, Magisterarbeiten usw.

für die kostenlose Publikation als Fachbuch.

Sie verfügen über eine Arbeit, die hohen inhaltlichen und formalen Ansprüchen genügt, und haben Interesse an einer honorarvergüteten Publikation?

Dann senden Sie bitte erste Informationen über sich und Ihre Arbeit per Email an *info@vdm-vsg.de*.

Sie erhalten kurzfristig unser Feedback!

VDM Verlagsservicegesellschaft mbH
Dudweiler Landstr. 99   Telefon +49 681 3720 174
D - 66123 Saarbrücken   Fax   +49 681 3720 1749
www.vdm-vsg.de

Die VDM Verlagsservicegesellschaft mbH vertritt

Printed by Books on Demand GmbH, Norderstedt / Germany